Basic Structures of Reality

BASIC STRUCTURES OF REALITY
ESSAYS IN META-PHYSICS

Colin McGinn

Oxford University Press, Inc., publishes works that further
Oxford University's objective of excellence
in research, scholarship, and education.

Oxford   New York
Auckland   Cape Town   Dar es Salaam   Hong Kong   Karachi
Kuala Lumpur   Madrid   Melbourne   Mexico City   Nairobi
New Delhi   Shanghai   Taipei   Toronto

With offices in
Argentina   Austria   Brazil   Chile   Czech Republic   France   Greece
Guatemala   Hungary   Italy   Japan   Poland   Portugal   Singapore
South Korea   Switzerland   Thailand   Turkey   Ukraine   Vietnam

Copyright © 2011 Oxford University Press

Published by Oxford University Press, Inc.
198 Madison Avenue, New York, New York 10016

www.oup.com

Oxford is a registered trademark of Oxford University Press

All rights reserved. No part of this publication may be reproduced,
stored in a retrieval system, or transmitted, in any form or by any means, electronic,
mechanical, photocopying, recording, or otherwise,
without the prior permission of Oxford University Press.

Library of Congress Cataloging-in-Publication Data
McGinn, Colin, 1950–
Basic structures of reality : essays in meta-physics / Colin McGinn.
  p. cm.
ISBN 978-0-19-984110-3 (alk. paper)
1. Physics—Philosophy. 2. Metaphysics. I. Title.
QC6.M315   2011
530.01—dc22   2011009272

Contents

Preface   vii

Part One

Introduction: Philosophy and Physics   3
1. The Concept of Matter   12
    Postscript: Particles as Fields: An Objection   43
    Appendix 1: The Uniformity of Matter   48
    Appendix 2: Divisibility and Size   53
2. What Is a Physical Object?   58
3. The Possibility of Motion   74
4. Motion, Change, and Physics   96
5. The Law of Inertia   110
6. Mass, Gravity, and Motion   123
7. Electric Charge: A Case Study   128
8. Two Types of Science   142
9. The Ontology of Energy   165
10. Consciousness as a Form of Matter   175
11. Matter and Meaning   192

Part Two

Foreword to *Principia Metaphysica*   211
Principia Metaphysica   213

Index   231

Preface to *Basic Structures of Reality*

This book consists of a series of essays written over a ten-year period, all of them published here for the first time. They were not originally conceived as chapters in a book, but as independent works. I have knitted them together into book form—and they do form a natural unity—but each essay can be read independently of the others. This has necessitated some repetition, for which I hope to be forgiven. The essays deal with questions at the intersection of physics, ontology, epistemology, and philosophy of mind. I hesitate to describe the book as "philosophy of physics": it is better described as philosophy conducted in the *company* of physics. I might even call it "philosophical physics" by analogy with "philosophical psychology" (as opposed to "philosophy of psychology"). It aims to develop a philosophical understanding of basic physical concepts. It does not attempt to do metaphysics *by* doing physics; nor does it subject physics to a metaphysical critique. Rather, it seeks to articulate the philosophical content of physics: what it presupposes, its distinctive mode of theorizing, the kind of knowledge it generates, and its wider significance.

My interest in the subject has two main sources. First, I have had a longstanding interest in the science itself—from an amateur perspective. I think this has something to do with the way physics combines the speculative and the rigorous. Physics is mind-stretching and controversial, both historically and contemporarily, yet also mathematical and experimental: it is rigor mixed with intellectual soaring. Physics is the imagination rendered mathematical. I see an affinity with philosophy, which is also speculative yet rigorous: big mind-stretching questions treated with careful analysis and argument. Philosophy is wonder rendered logical. Philosophy and physics thus make natural partners. Second, physics attempts to plumb the depths of the world—it asks ultimate questions and seeks ultimate answers—and thus represents human inquiry at its most ambitious. It is therefore an ideal place to examine the scope and limits of human knowledge. Physics is a case study in epistemology. I have a special interest in that because of my interest in natural mysteries—phenomena of the world that baffle our theoretical efforts. How mysterious *is* the physical world? Are its mysteries related to other mysteries—such as the mystery of consciousness? What is the source of the mysteries of physics, if there are such? In physics we see the human mind at its outer limit—thrusting and striving, succeeding and failing. Physics casts its sharp light on the natural world, but it thereby highlights the vast darkness that surrounds the patches of illumination (or so I shall be arguing).

Part Two of the book, boldly entitled *Principia Metaphysica*, continues the theme of knowledge and its limits, combined with intellectual striving—the desire to reach a final comprehension of reality and the accompanying sense of bafflement and limitation. It mainly concerns laws of nature. As to form, the text is arranged as a series of aphorisms, in the style of Wittgenstein's *Tractatus*. For the first and only time in my intellectual life, a set of ideas presented themselves to me in the shape of philosophical aphorisms—almost as if I were taking dictation from an impatient and importunate alien. I decided not to fight it, though I did trim and tinker, expand and elucidate. I fondly view it as a philosophical poem, larded with metaphor, oblique and haughty—somewhat infuriating perhaps. Make of it what you can.

I mostly worked on these essays in solitude, not soliciting responses or presenting them publicly (though I did do a couple of sparsely attended seminars on the material here in Miami). However, I was the fortunate beneficiary of continuous discussions with my colleague Peter Lewis, who both criticized my ideas and reassured me: his open-mindedness and calm skepticism were invaluable. I am also grateful to Otavio Bueno for his receptiveness and probing questions. Galen Strawson made many incisive and insightful comments on the text I submitted to the publisher. All three amply earned the appellation *colleague*.

Colin McGinn

December 2010
Miami

PART ONE

Introduction
Philosophy and Physics

Some years ago I decided to turn my attention to fundamental metaphysics. I wanted to think about the most basic categories of reality. Objects in space seemed like a good place to start. This led me to physics—the science of the motions of matter in space. What exactly is matter? How is it related to space? How is motion to be understood? Philosophers have discussed these questions, as have physicists; indeed, the distinction between these two types of enquirer hardly exists where such questions are concerned. They are questions of "natural philosophy." I began studying physics and its philosophy, finding it quite stimulating. But it wasn't easy: I had no real background in physics as a science and hadn't seriously worked on philosophy of physics before. I had to learn it from scratch. I approached the subject not as a narrow specialist, but as a general philosopher interested in the most general questions. I gave some seminars and wrote some exploratory papers, and eventually I accumulated quite a bit of material, unified around certain central themes. This book is based on that material. In this introduction, I want to spell out what those themes are in a preliminary way and outline my general orientation.

Recently, I came across the following statement from Michael Dummett in his *The Nature and Future of Philosophy*:

The greatest lack, however, is of philosophers equipped to handle questions arising from modern physics [or older physics!]; very few know anything like enough physics to be able to do so. This is a serious defect, because modern physical theories impinge profoundly upon deep metaphysical questions it is the business of philosophy to answer. We may hope that some philosophers may become sufficiently aware of this lack to acquire a knowledge of physics adequate for them both to integrate it with their treatment of metaphysical problems and to convey to philosophical colleagues who know less physics what they are talking about. (150)[1]

This describes my attitude perfectly, however imperfect my execution of it may be. I have tried to acquire such knowledge of physics as will serve in areas conceptually adjacent to physics—epistemology, metaphysics, philosophy of mind, even theory of meaning. I also take seriously the injunction to be clear to my colleagues who may not know much about the details of physical theories: nothing I say in this book should be baffling to readers innocent of the technicalities—no equations, no unexplained jargon, no mystification. The book consists of philosophical reflections on basic physics, accessible without grasp of technical and mathematical subtleties (it will probably disappoint *real* philosophers of physics). Instead of practicing philosophy of physics as an arcane specialism, I want to unify it with philosophy as a whole; I am in the business of crossing lines.

To be concrete: there is very little here on those staples of contemporary philosophy of physics, quantum theory and relativity theory—certainly nothing particularly exacting or esoteric. The issues I discuss can be raised independently of those parts of physics; it is not that physics became epistemologically problematic only with the advent of twentieth century physics. Returning to Newton would not remove all our perplexities. So we need not enter into the details of contemporary physics, troubling as they are, in order to grasp the basic philosophical problems I want to discuss. Quantum theory certainly adds to the perplexity, but it is not the sole source of it. I am concerned only with the most general categories of physical theory. Thus I am not so much interested in doing philosophy *of* physics, as it now exists as a developed science, but in asking questions of what might be called "natural metaphysics"—but in the presence of the science called "physics" as it has formed over the centuries. In the time of the great physicists of the seventeenth and eighteenth centuries—Galileo, Descartes, Newton, et al—physics was closely integrated into general philosophy, with Locke, Leibniz, Berkeley, and Hume fully cognizant of what was going on. I am trying to go back to that earlier time, in spirit and style. So this book is not intended only for specialists.

Two concerns dominate the essays that follow: what kind of knowledge of reality physics affords, and what is the nature of the basic ontology of physics. The epistemological question has been amply discussed, early and late, and I make no pretensions to originality in advocating the so-called "structuralist" conception of physical

---

[1] Dummett, *The Nature and Future of Philosophy* (Columbia University Press: New York, 2010). Dummett is discussing the current situation in philosophy and noting some flaws and problems.

knowledge. Who exactly originated this view is an interesting historical question, but traces of it can be found in Hertz[2] and Poincare, and explicit statements abound in Russell and Eddington. Roughly speaking, it is the view that our knowledge of physics does not disclose the intrinsic nature of the entities postulated, but only their mathematically specified interrelations. The knowledge is "abstract," leaving it an open question what physical reality is really like in itself. In other words, a deep ignorance lies at the heart of physics, despite its formal richness. I expound this position at several points in the essays that follow, so I will not say more now; in any case, the general shape of the view is quite well known. I shall be concerned mainly with its significance and interpretation; I take the view to have been convincingly established by others.[3]

It is in part because of the attractions of the structuralist position that instrumentalism and conventionalism have been found appealing with respect to physical theories: for if we cannot really come to know what physical reality is like in itself, are not our theories merely conventionally adopted instruments for predicting what we *can* know, viz. observations? Structuralism thus leads easily to positivism. And positivism has dogged physics ever since the epistemological limits of physics began to become apparent (I would date that to the time of Newton).[4] One of my questions will be how to avoid positivism while accepting structuralism (short answer:

---

[2] In *The Principles of Mechanics* (Dover Publications: New York, 1956) Hertz writes: "We form for ourselves images or symbols of external objects; and the form which we give them is such that the necessary consequents of the images in thought are always the images of the necessary consequents in nature of the things pictured. In order that this requirement may be satisfied, there must be a certain conformity between nature and our thought. Experience teaches us that the requirement can be satisfied, and hence that such a conformity does in fact exist.... The images which we here speak of are our conceptions of things. With the things themselves they are in conformity in *one* important respect, namely, in satisfying the above-mentioned requirement. For our purpose it is not necessary that they should be in conformity with the things in any other respect whatever. As a matter of fact, we do not know, nor have we any means of knowing, whether our conceptions of things are in conformity with them in any other than this *one* fundamental respect"(1–2). This is quite a clear early statement of the structuralist position. I discuss the other authors mentioned in the text in the course of this book. (Note that the Wittgenstein of the *Tractatus* had read Hertz: the picture theory of meaning is strongly suggested by Hertz's words.)

[3] For a survey of structuralist thinking over the years, see Mark Lange's article "Structural Realism" in the online *Stanford Encyclopedia of Philosophy* (2009).

[4] In a fascinating passage at the very end of *Principia* (1726) Newton writes: "As a blind man has no ideas of colours, so have we no idea of the manner by which the all-wise God perceives and understands all things. He is utterly void of all body and bodily figure, and can therefore neither be seen, nor heard, nor touched; nor ought he to be worshipped under the representation of any corporeal being. We have ideas of his attributes, but what the real substance of any thing is we know not. In bodies, we see only their figures and colours, we hear only the sounds, we touch only their outward surfaces, we smell only the smells, and taste the savours; but their inward substances are not to be known either by our senses, or by any reflex act of our minds: much less, then, have we any idea of the substance of God" (441–42). He immediately goes on to concede that he has no idea of the cause of gravity, as distinct from its laws, famously remarking that he "frames no hypotheses." This frank admission of epistemological limits had an incalculable impact on how physics was subsequently understood.

extreme principled agnostic realism). In addition to this question of the scope and limits of physical knowledge, there is also the question of the status of basic propositions of physics, such as Newton's laws of motion: are they empirical generalizations or a priori principles or something else? What about the conservation of energy? Is the traditional division between the a priori and the a posteriori even sustainable in the light of these kinds of propositions? Where does a priori metaphysics end and empirical physics begin?

With respect to the ontological question, the problems are immediate and massive. What is the nature of space, time, matter, motion, force, and energy? These have all been hotly contested in the history of the subject, despite their foundational role in actual physical theories. Even classical mechanics was plagued by such ontological questions. I consider these questions in some depth in the following essays; and my main mission is to avoid verificationist and operationalist dismissals of very real issues. In general, I defend realist interpretations of these basic categories, even when it turns out that our concepts are quite inadequate to capture what is really going on out there. Our concept of motion, in particular, is very difficult to make sense of, and the very possibility of motion-as-we-conceive-it can be cast into serious doubt (Zeno lives!). The relation between space and matter is highly contentious. Matter itself is vanishingly elusive. Electric charge is utterly mysterious. Gravity baffles. The mathematics is as regular as clockwork, but the reality to which it applies defies our best efforts at comprehension. It is not that we see reality as through a glass darkly; we don't see it at all, but we do have a marvelously precise mirage that can take its place. Physics does not give us a dim or distorted view of reality; it gives us a perfectly clear view of reality under a mathematical description—but only that. The "ontological commitments" of physics are obscure, because we have so slight a grip on the real nature of what we postulate. And even when we think we know what we are talking about (as with space, matter, and motion) deep conceptual problems confront us. The reason there is—and always has been—a thriving philosophy of physics is that physics inherently raises hard philosophical questions that refuse to go gently into that good night. What, most primitively, is physics really *about*?

I come to physics from a philosophy of mind background. That background is never far from my thinking about physics. I hold that the mind, particularly consciousness, is a natural mystery: we don't, and probably can't, understand how the conscious mind fits into nature as a whole, specifically the part we call the brain.[5] Looking back to my early formulations of this thesis, I sense some naivety on my part about physics: I seemed to assume that physics was free of deep mystery, that physical theories are fully intelligible (of course, I made an exception for quantum theory). Newtonian mechanics, I assumed, was not steeped in the mystery that surrounds consciousness; chemistry was in better shape than neuroscience. Well, it may

---

[5] I discuss this in a number of works, including *The Problem of Consciousness* (Basil Blackwell: Oxford, 1991) and *Consciousness and Its Objects* (Oxford University Press: 2004).

be true that the physical world is not *as* mysterious as consciousness, but that doesn't mean it carries *no* burden of mystery. It is not all sweetness and light in physics and gloom and frustration in consciousness studies. In short, there are "mysteries of nature" in physics as well as psychology.[6]

Actually, the entire world is shot through with mystery, as it now seems to me—in different degrees, perhaps, and for different reasons, but with no safe oasis (maybe mathematics, maybe ethics). This recognition changes one's view of the mysteries of mind. The mind does not now seem so *special* in its mysteriousness: it is not like a lone dark island on an otherwise well-illuminated landscape. We don't any longer think that the mind is *uniquely* blocked from human comprehension—so it is not to be regarded as an anomalous exception to nature's general epistemological obligingness. Its mysterious character begins to seem like the rule not the exception: ignorance is widespread and apparently incurable, even in the hardest of sciences. Physics too has its "hard" problems, as well as its "easy" problems. If nature as a whole, including the simple motions of matter, is shrouded in mystery, then we shouldn't be amazed to discover that mind is also. And, as physics shows us, mystery is compatible with sound science: physics proceeds pretty nicely as a science without managing to resolve its deep ontological mysteries satisfactorily (without even having a solid and uncontroversial concept of motion).

Analogously, we could have a thoroughly decent science of consciousness without solving the mind-body problem. The structuralist position enables us to see why: precise predictive knowledge, mathematically formulated, is consistent with utter ignorance of the intrinsic character of the subject matter of that knowledge. So I can look forward to a flourishing science of the mind even as I proclaim that deep ontological questions will remain unresolved—just as we have seen in the course of the development of physics. The case of physics teaches us that mystery is not an impediment to science; indeed, one might be forgiven for supposing that mystery is a *necessary condition* for successful science![7] In any case, I now think that

---

[6] The phrase is Hume's, unearthed by Noam Chomsky in his "Mysteries of Nature: How Deeply Hidden?" *Journal of Philosophy* 106, no. 4 (April 2009). I have, in general, been strongly influenced by this article and other writings by Chomsky on natural mysteries and cognitive limits. The passage from Hume's *History of England* (1754–62) is so magnificent that it is worth quoting in full: "In Newton this island may boast of having produced the greatest and rarest genius that ever rose for the ornament and instruction of the species. Cautious in admitting no principles but such as were founded on experiment; but resolute to adopt every such principle, however new or unusual: from modesty, ignorant of his superiority above the rest of mankind; and thence less careful to accommodate his reasonings to common apprehensions: more anxious to merit than to acquire fame: he was, from these causes, long unknown to the world; but his reputation at last broke out with a lustre which scarcely any writer, during his own lifetime, had ever before attained. While Newton seemed to draw off the veil from some of the mysteries of nature, he showed at the same time the imperfections of the mechanical philosophy; and thereby restored her ultimate secrets to that obscurity in which they ever did and ever will remain. He died in 1727, aged 85" (328–29).

[7] The science needs to be able to select a tractable aspect of its domain for treatment, ignoring the rest. Physics selects formal aspects of its domain, these yielding to mathematical treatment; the

consciousness is on a continuum of mysteries, beginning even with simple mechanics, not a splendidly isolated enigma (perhaps, though, it still qualifies as the *biggest* mystery). Human knowledge has its gaps and blind spots all over the place (as well as its high points and triumphs).[8]

The philosopher of mind with an interest in physics may also wonder whether "physicalism" or "materialism" is a doctrine to defend. Certainly a doctrine so named *has* been defended. Let me put it firmly on record that I am opposed to any such doctrine, though I am often described as a "physicalist mysterian." I take up this question at different points in the book, generally arguing that the alleged doctrine of "physicalism" is ill defined because there is no workable notion of the "physical" to appeal to. Here I follow closely Chomsky's position: since the demise of mechanism, with the advent of Newtonian action-at-a-distance, there is no clear notion of the "physical."[9] I believe in the *dis*unity of physics: physics deals with disparate ontological categories with no unifying conceptual framework—regions and points of space, instants and durations of time, fields of force, causation through vacuum, accelerating bodies, electrons, protons and other particles (some with mass and some without, some charged, some not), light rays, units of energy, waves, curved space-time, strings, multiple universes, singularities, dark matter, and so on. It is all a far cry from Descartes' tidy mechanics of extended substance interacting by contact causation. I won't rehearse the usual arguments here, but my general position is that it is either false or empty to say that the mental reduces to the physical, depending on what we choose to mean by the elastic word "physical."[10]

What is far more interesting and substantive is the question of how physics itself should be formulated: what properties, processes, and entities it should canonically postulate. That is a question about the proper resources of physics, about what is eliminable in favor of what, and what might be reducible to what else. Different

---

rest it leaves mysterious. What is noteworthy is that the science is successful proportionately to what it leaves out—as if it works so well *because* of what it leaves out. If the science of matter could really delve into the intrinsic nature of its domain, it might not have the formal rigor it purchases as the price of its selectivity.

[8] It would be illuminating to develop a taxonomy of mysteries, with a hierarchical structure: from the most elementary mysteries to the most sophisticated. The mystery of matter as such might lie on the lowest level, with consciousness right on the top. The special characteristics of consciousness do seem to put it in a class apart (perhaps right next to free will and creativity). We might accordingly distinguish "regular mysteries" from "super-mysteries." It is certainly a bad argument to suggest that consciousness can't be such a big mystery if even successful physics has its mysteries: for mysteries might vary in their depth and magnitude, and mystery is still mystery.

[9] I discuss the lack of a decent definition of "physicalism" in *Consciousness and Its Objects* (16ff), cited in note 5. It is a longstanding theme of Chomsky's.

[10] This is often referred to as "Hempel's Dilemma": either by "physical" we mean what is recognized in current physics, which is too narrow; or we mean what would feature in ideal physics, in which case it is trivial (since that would simply be the true description of the world). The underlying problem is that people pick a few paradigms to pin down what they mean by "physical" and then say (in effect) "and anything like *those* things": but no workable standard of similarity is proposed. Thus are pseudo-concepts let loose.

degrees of austerity have historically characterized these debates. Cartesian mechanism was highly austere—allowing only extended bodies and causation by contact. Newtonian mechanics allowed in action at a distance, which seemed like mysticism to the Cartesians. The ether was wheeled in to ease the sense of the occult, regaining austerity in letter but surely not spirit. Hertz constructed a mechanics without Newtonian force, deeming this too metaphysical for his taste. Maxwell's electromagnetic fields added greater laxity, with or without the intervening ether. Nowadays we have quite a liberal physics: several types of force operating through the vacuum at arbitrary distances, many types of elementary particle, electric charges, fields, dark matter, multiple universes, nonlocality, and so on. You can well imagine an old-time physicist protesting that this stuff is just not *physical*—meaning, solid bodies colliding in space. Such a reactionary physicist would be insisting that modern physics has gone beyond the bounds of the "physical"; and he or she might even attempt to demonstrate that some contemporary constructs can be reduced to physical properties in the old sense (this is precisely what the ether was supposed to do with respect to mechanism). But the point is that these wrangles about "physicalism" are pointless and distracting. So you won't find me arguing for, or even against, "physicalism" (or "materialism") in this book, at least as recent philosophers have thought to understand these terms, since I don't think the doctrine is well defined or even very interesting.[11]

In the essays that follow certain themes recur, which may as well be highlighted now; this may help guide the reader through the thickets. The troubled conceptual history of physics comes up a lot: we cannot suppose that physics progressed smoothly and incrementally from the elementary to the advanced, with agreement on the fundamentals—there were always deep metaphysical and epistemological issues at play (as there still are today). This is why physics has always been close to philosophy—while botany, say, has not. I would describe the history of physics as philosophically *vexing*, with continual stresses and strains on the human conceptual scheme; *violence* would not be too strong a word. A profound sense of conceptual inadequacy or insufficiency seems endemic and ineradicable. The foundations always seem in question, even as the day-to-day science flourishes. Because of this philosophical uncertainty, there is a standing temptation to try to reduce physics to its least controversial part, namely observations. The positivism of Mach is just the most extreme version of this: the sole content of physical theories is *sensations*—all the rest is artifice.[12] Twentieth-century physics developed in tandem with the popularity of logical positivism; and traces of positivism are quite discernible in the writings of both Einstein and Heisenberg, to name just two. Since I believe that positivism has long since been refuted, I am anxious that physics not bear the stamp of it; so one of my standing concerns is the legacy of positivism, especially verificationism. No argument will here be regarded as persuasive that simply presupposes verificationism (arguments against absolute

---

[11] I discuss this most fully in chapter 10 of this book.
[12] The *locus classicus* is Ernst Mach's *The Analysis of Sensations* (La Salle: Open Court, 1984).

motion are often blatantly verificationist). We need to disentangle the physics proper from this brand of philosophical theorizing.

The puzzling and surprising will not go unremarked. It is all too easy to acquiesce in the findings of physics as just one more fact to be added to the others (the textbooks tend to give this impression). A philosopher needs to have a nose for what is conceptually startling or problematic. Gravity is a very surprising thing, despite the familiarity of the concept, not just because of the remote causation it entails, but also by linking mass and attractive force in the way it does. Electricity is doubly surprising, because of its repulsive *and* attractive powers. The constancy of the speed of light is absolutely astonishing. The transmutability of energy is strikingly strange, almost alchemical. I discuss each of these puzzling phenomena, accentuating their oddity. The mathematics should not blind us to the sheer peculiarity of the physical world (and this is before we get to relativity theory and quantum theory). And then there is the meta-question: *why* do we find the world so odd? The natural answer is: because it departs from common sense. But why is common sense so out of harmony with objective reality? Why should common sense have *any* views about the very small or the very large or the velocity of light in a vacuum? Or could it be that the reason nature seems so odd to us is that we haven't really got to the bottom of it? Does God find it odd? To what extent *should* the world be intelligible? Whatever the case may be, we do well to maintain a sharp sense of the surprising and counterintuitive.

Physics has endured so many crises, with corresponding revolutions, and is still today in such turmoil, that it is hard to believe it will not need radical change in the future. Some of what is needed may be philosophical, as well as empirical. I think a strong sense of its limitations, as a body of knowledge purporting to be about a reality existing beyond human thought, is essential to appreciating where its weaknesses may lie. The mathematical method, though it has yielded great insights, has not allowed us to penetrate to physical reality as it is in itself—that is the central point of the structuralist perspective. In physics we see the magnificence of the human intellect, but also its profound limitations. Our picture of the physical world is obviously a product of the method we use to comprehend it—the formulation of laws in mathematical terms. There is no guarantee that this method will yield complete knowledge of how the world is constituted objectively. Our knowledge is glancing and in a sense superficial—a mapping of interdependencies between things we only dimly grasp, if at all. Physics has been described as providing an "absolute conception" of the world, in the sense that it abstracts away from the specific subjective viewpoint.[13] That sounds right, but it doesn't follow that it is an "intrinsic conception" of the world, i.e., a conception that reveals what the world is really like in itself. The mathematical nature of physics serves to make it absolute, because mathematics is far removed from human subjectivity, but

---

[13] The phrase is Bernard Williams' from *Descartes: The Project of Pure Enquiry* (Harvester Press: London, 1978). I discuss the topic in *The Subjective View* (Oxford University Press: Oxford, 1982). See also Thomas Nagel, *The View from Nowhere* (Oxford University Press: New York, 1986).

its very mathematical character condemns it to abstractness—to being merely "structural."¹⁴ And if physics does not give us an intrinsic conception, it is unlikely to give us a properly intelligible conception, since the world is presumably intelligible only from its own point of view, i.e., when viewed intrinsically. Such, at any rate, will be the position developed in the course of the essays that follow.

Aside from that polemical purpose, the book aims to provide one particular philosopher's attempt to integrate physics with some broader philosophical questions. What can physics teach us about epistemology, metaphysics, and philosophy of mind? And what can those disciplines teach the reflective physicist? In answering these questions, I want to give physics its due, but not to be intimidated by it. No doubt "physics envy" is an understandable malady for a philosopher (among others), but at the same time providing a just estimate of the limitations of physics is a natural philosophical task. The emperor is not short of a suit of fine clothes, to be sure, but whether he has a substantial body underneath is the more serious question (is he nothing but structure?). With all the instruments and measurements and calculations and equations, not to speak of the contributions of human genius, it is still not clear that physics has really revealed what nature is ultimately made of and how it fundamentally works. It has given us a shadow of nature, a skeleton maybe, clothed in dazzling mathematical robes, but nature's body still lies essentially hidden—maybe necessarily so. We ourselves are ultimately products of the physical world—effects of the big bang, in fact—and whether we can attain an objectively accurate and complete grasp of the reality that produced us is very much an open question.¹⁵

---

¹⁴ It is arguable that the mathematical character of physics, as we have it, reflects a human contribution to its content, because mathematics is *our* preferred way of understanding the world. But is the external world mathematical in its inherent nature? Does God view the physical world mathematically? We seem to be imposing a mathematical grid on it, as a condition of human comprehension. Perhaps Martian physicists would impose the same grid—but it might still be a grid. Certainly, the scales of measurement we use reflect our choices. Does it really make sense to say that the physical world is objectively mathematical—quite independently of any system of representation an intelligent enquirer might bring to the world? Is mathematics *intrinsic* to motion (say)? Mathematics seems extrinsic to the physical subject matter. (None of this is to dispute the objectivity and truth of physics.)

¹⁵ As has often been remarked, it is surprising that we know as much as we do, given our origins and the forces that shaped human intelligence. We are evolved organisms, with specialized finite brains, not all-purpose cognitive gods. Why then is human intelligence so often assumed to be radically discontinuous with that of our animal kin, including those species closest to us? Why are other animals thought to be naturally ignorant, while we are taken to be naturally all knowing? Human scientific knowledge, to a properly naturalistic perspective, is really a patched-together, shakily based, mish-mash of the empirical and formal—highly inferential and local, laboriously acquired, extremely fallible, and contrary to our natural tendencies. The science-forming faculty (as Chomsky calls it) is like our mountain-climbing faculty: not part of what nature intended and operating only by tremendous effort and artifice—and not all peaks are guaranteed to be attainable. There is nothing *automatic* about scientific knowledge, nothing assured or promised or inevitable or innate (unlike, plausibly, knowledge of language or folk psychology). What we know of the universe beyond is the result of a kind of dumb biological luck, with no guarantees built into it. See my *Problems in Philosophy: The Limits of Enquiry* (Basil Blackwell: Oxford, 1993) for more on the bounds of human thought.

1 The Concept of Matter

DESCARTES AND LOCKE ON EXTENSION AND IMPENETRABILITY

In an earlier epoch, the problem of defining matter was taken very seriously. No one took the concept for granted. Without an adequate concept of matter, physics was felt to lack coherence—a clear subject matter. (In the twentieth century the concept of mind was similarly contentious.) In the *Discourse on Method*[1] Descartes writes as follows: "the nature of matter or body, considered in general, does not consist in its being hard, or ponderous, or colored, or that which affects our senses in any other way, but simply in its being a substance extended in length, breadth, and depth" (198). In *The Principles of Philosophy*[2] he says: "Extension in length, width and breadth constitutes the nature of physical substance, and thought constitutes the nature of thinking substance. For everything else that can be attributed to a body presupposes extension, and is merely a certain mode of an extended thing; likewise all the things that we find in the mind are merely different modes of thinking. Thus, for example, we cannot understand shape except in an extended thing, or motion except in an extended space; likewise, we cannot understand imagination, or sensation or willing except in a thinking thing" (132). In the

---

[1] I use the edition by Barnes and Noble Books, 2004.
[2] References are to the Penguin Classics *Meditations and Other Metaphysical Writings* (Penguin Books: London, 1998).

Meditations[3] he strikes a somewhat different note: "by a body I understand anything that can be limited by some shape, can be circumscribed in a place, and can so fill a space that every other object is excluded from it" (25). That last clause is curious in the light of remarks he makes in his reply to Henry More[4], where he is adamant that impenetrability is not part of the definition of matter: "tangibility and impenetrability in a body are similar to the ability to laugh in the case of human beings, a property of the fourth type according to the common rules of logic and not a genuine, essential difference, whereas I contend that extension is such an essential difference. Therefore, just as human beings are not defined as animals capable of laughter but as rational animals, so likewise body is not defined by impenetrability but by extension" (168). In any case, his official definition of matter is in terms of pure extension, not exclusion or impenetrability.

Descartes was well aware that space might be similarly defined, since even empty space (so called) is extended in three dimensions. He does not count this as an objection, however, since, according to him, matter and space are not fundamentally different, space being a type of physical substance. In the *Discourse* he writes: "Space or internal place, and the corporeal substance that is comprised in it, are not different in reality, but merely in the mode in which they are wont to be conceived by us. For, in truth, the same extension in length, breadth, and depth, which constitutes space, constitutes body" (201). He also remarks: "After this examination we find that nothing remains in the idea of body, except that it is something extended in length, breadth, and depth: and this something is comprised in our idea of space, not only of that which is full of body, but even of what is called void space" (202). This doctrine is part of his opposition to the notion of a vacuum: for him, space is not emptiness but a "plenum," a type of substance in its own right (though intangible and invisible). Perhaps, indeed, the doctrine of the plenum makes the thesis of impenetrability look shaky, since it appears that body and space can interpenetrate—and both are types of physical substance, according to Descartes. Matter does not exclude space, apparently, and yet space is a type of matter (in effect), since it is defined by extension. So *some* matter (the spatial kind) is not impenetrable. In any case, Descartes is ready to admit that space and matter share the same defining property: extendedness. It is important to note that, for Descartes, matter is not defined as what *occupies* extended space; it is defined by extension itself. Clearly, impenetrable objects occupy space and are extended in it, but Descartes is saying something stronger—to be a material body simply *is* to have the property of extension (which is why space itself also qualifies). Space is penetrable, he concedes, but it has extension, so it meets his condition for being material. Merely having *extent* is sufficient for materiality—measurability in three dimensions (presumably he does not think that extension in two dimensions is sufficient). This is what is *intrinsic* to the concept

---

[3] See the same volume as in note 2.
[4] Again, see the same volume as in note 2.

of matter, while impenetrability is at best extrinsic—not part of its very constitutive essence.

Locke was strongly opposed to Descartes' definition of matter. In "Of Solidity" in his *Essay*[5] he is clearly disagreeing with Descartes in placing extension at the heart of the concept of matter, instead of impenetrability. He writes: "That which thus hinders the approach of two bodies, when they are moving one towards another, I call *solidity*. I will not dispute, whether this acceptation of the word *solid* be nearer to its original signification, than that which mathematicians use it in…but if anyone think it better to call it *impenetrability*, he has my consent. Only I have thought the term *solidity*, the more proper to express this idea, not only because of its vulgar use in that sense; but also, because it carries something more of positive in it, than *impenetrability*, which is negative, and is, perhaps, more a consequence of *solidity*, than *solidity* itself. This of all other, seems the idea most intimately connected with, and essential to body, so as nowhere else to be found or imagined, but only in matter" (125). Naturally, then, he sharply distinguishes matter from space, whose essence is penetrability. He says: "This resistance [of matter], whereby it keeps other bodies out of the space which it possesses, is so great, that no force, how great soever, can surmount it. All the bodies in the world, pressing a drop of water on all sides, will never be able to overcome the resistance which it will make, as soft as it is, to their approaching one another, till it be removed out of their way; whereby our idea of solidity is distinguished both from pure space, which is capable neither of resistance nor motion, and from the ordinary idea of hardness."(125) Mere extension is common to space and matter: what distinguishes them is the penetrability of the one and the impenetrability of the other. He thus rejects Descartes' notion of the plenum, maintaining that space in itself is a pure vacuum. Motion would not be possible without such a vacuum, he thinks, since matter is essentially resistant to motion. He also carefully distinguishes solidity, in his sense, from hardness: a hard object is one whose internal parts are tightly bound to one another, unlike a soft object, but all objects are equally solid. He writes: "But this difficulty of changing the situation of the sensible parts amongst themselves, or of the figure of the whole, gives no more solidity to the hardest body in the world, than to the softest; nor is adamant one jot more solid than water" (126). Thus every piece of matter is as exclusive as any other, even liquids and gases; the differences in resistance are wholly a matter of the forces that bind the constituent particles together. He gives the example of air in a football: if the constituents of the volume of air are prevented from moving sideways, the football will be as resistant as the hardest steel. Solidity is an absolute quality of material things; it does not come in degrees. Matter is defined by solidity. For an object to occupy a region of space is for its solidity to exclude other objects from that region.

Now I basically agree with Locke as against Descartes about the difference between matter and space, but at this point I propose to drop matters of history and exegesis. I want now to give my own sketch of the viewpoint represented by Locke, for later

---

[5] *An Essay Concerning Human Understanding* (Penguin Books: London, 1997).

use in developing a theory of matter and space that goes well beyond his treatment, as follows. The physical universe consists of matter in space. Matter is characterized by impenetrability, which is equal and constant from one material body to another. Space is a vacuum and is defined by penetrability, which is also equal and constant (no part of space is more receptive to occupation than any other part). Extension is common to both space and matter and so cannot be used to define either. Material bodies consist of moveable parts. Space consists of immoveable parts. Motion is the passage of impenetrable bodies through penetrable space. There cannot be motion through matter of any kind, no matter how "rarified." Motion requires a vacuum because all matter is resistant; yet particles in space can be pushed aside by moving objects. Matter is compressible, but only because particles can be made to approach nearer to each other in the interior space of the object. Continuous matter would all be equally hard, since hardness or softness depends upon the movement of particles within the object. The particles themselves cannot differ in their degree of hardness. Thus there are two types of resistance that objects can exhibit: the kind that results from the cohesion of parts, which is a matter of forces; and the kind that results from matter as such, which cannot come in degrees. The second kind might be called logical or metaphysical, for want of other terms; the former is a matter of natural law. Matter and space are logical or metaphysical opposites, one defined by exclusiveness, the other by receptiveness. It is a mistake to think of space as if it were a kind of exceptionally soft matter. Whether space can be properly described as a "substance" is an ill-defined question, pending some clear idea of what a substance is. What is important is that space and matter differ essentially. This is no doubt all rather simple and crude, and dogmatically stated, but I think it will serve my purposes in what follows.

A PROBLEM FOR THE LOCKEAN VIEW

It is natural to object to impenetrability on the ground that this is a dispositional or relational concept, not an intrinsic one: it is the ability to hinder movement in another object. Locke himself admits that impenetrability is more a consequence of solidity (i.e., spatial occupancy) than constitutive of it. One wants to ask in virtue of *what* matter is impenetrable. We can't say that it is in virtue of being *matter*, because that is both circular and an admission that the two properties are not the same. The intuitive idea is that of occupying, or taking up, space—of what Locke calls "repletion." It is *because* matter takes up space, possesses it, that it is impenetrable. But what is this idea of occupying space? It cannot just be excluding other objects, or else it couldn't *ground* the disposition of impenetrability. What matter does is *fill* space, and in so doing it excludes other matter from occupying the space filled. This appears to be what Locke intends by the word "solidity," since he regards this as the *basis* of the relation of impenetrability. Perhaps, too, Descartes is working with this intuitive idea and expressing it by means of the idea of extension *in* space, as opposed to extension as such. The taking up of a certain region of space certainly seems to capture the right idea, once the taking up in question is assumed to entail exclusion. So the most basic notion here seems to be spatial occupancy, which generates both extensiveness and exclusiveness. But what account are we to give of this notion of occupancy? What exactly *is* it to occupy space, if this notion is not captured by either extension or impenetrability? That is the key question I shall be trying to answer in this paper: what is it for a body to *take up* space? What exactly does matter *do* to space when it takes it up? It renders it impenetrable, we know that—but in virtue of what does it do that? *How* does the exclusion work? Not by means of repulsive forces, because, as we have seen, the Lockean notion of impenetrability is not the same as hardness: it comes from the very nature of matter itself, not the forces that govern it. The exclusion is logical or metaphysical, not causal—not like the difficulty of making a dent in steel.[6] What is the *explanation* of matter's inherent impenetrability?

There is also a prima facie problem with this notion of occupancy as definitive of matter and the basis of impenetrability, arising from the notion of *force*. Do not fields of force pervade or permeate regions of space, and hence occupy them? Take the magnetic field of the earth: it appears to occupy a certain specific region of space, to act within a certain area. It has Cartesian extension. But surely a force field is different from a body—it is not a kind of matter (this is why electromagnetic force was such issue when it was first introduced). Gravitational field is the same: it appears to pervade space, and yet it is not a material object—so spatial occupancy cannot define the latter concept. If this is doubted, then observe that such fields are not impenetrable: distinct fields *can* pervade the same region of space—as gravitational and magnetic fields both surround the earth. They do not exclude each other, unlike volumes of matter. If spatial occupancy is to be the key to the nature of matter, where this produces impenetrability, then we need to explain how force fields and matter differ in their relation to space. This seems to me quite a difficult question, and I don't think we have very clear ideas of how fields and space relate, but there are some intuitive distinctions we can point to. It doesn't quite ring true to speak of fields as "taking up" space, or even "occupying" it; it is more that they "pervade" or "permeate" space. What this appears to mean, roughly, is that they operate within a

---

[6] This Lockean conception of impenetrability seems not to be generally accepted by physicists, who tend to speak as if only resistance by forces produces exclusion. But there is no good argument for this—essentially metaphysical—position: no empirical fact of physics compels us to accept that this is all that impenetrability amounts to. I myself think that the exclusion thesis is a piece of pure modal metaphysics, like the claim that nothing can be red and green all over at the same time—that is, an example of metaphysical necessity. It is simply part of the *essence* of matter to exclude other matter (as it is also part of the essence of matter to be located in space). Not every truth about matter is discovered by empirical physics or reflects metaphysically contingent laws. (The postscript following this paper discusses the metaphysical basis of impenetrability further, disputing the reduction of material exclusion to the existence of a repulsive force field. I largely take exclusion for granted in the body of the paper, focusing on what might explain it.)

certain region of space: that is, objects within that region are affected in certain ways by the force field in question. Hence the tendency to want to explain fields in terms of conditionals about objects: For there to be a magnetic field of a certain intensity at a certain point in space is for objects to be attracted or repelled in certain ways, as if the existence of the field reduces to the behavior of objects: if the object is placed at a certain point, will behave thus and so. No doubt this is too reductive, but it indicates that we don't think of fields as existing independently of objects—that somehow their nature consists in the effect they have on objects. They are not objects in their own right, but influences *on* objects. One wants to say that they act within certain regions of space instead of existing in those regions. The sense in which they are *in* space is obscure. They don't take up space in the exclusionary sense, because they are not fully occupants of space. We might want to distinguish weak and strong spatial location: the pervading kind and the occupying kind. A field of force has an extension, and is indeed extended in space, but it doesn't stake a claim on space in the way matter does—it is willing to share the space it "occupies." Fields are not like gases; they are not a rarified form of matter any more than space is. They seem to be midway between space and matter, neither quite one nor the other. Space is extended; fields are pervasive; and matter is exclusive. Fields evidently don't do to space what matter does when it renders a place impenetrable by other matter. A mark of this difference is that fields fall off in strength over distance—they become weaker in remoter regions of space. But material bodies don't do this; each part of the space they occupy is equally impenetrable. Fields are *less* pervasive or present in some areas than others, but a piece of matter does not occupy space any less at any point within it—certainly not towards its outer boundaries. The mode of "occupancy" is quite different in the two cases. Matter occupies space in an all-or-nothing way, but fields pervade space to one degree or another, allowing other fields to coexist with them. We shouldn't then assimilate the spatiality of fields to that of material bodies. There is also the point, for what it is worth, that fields, unlike matter, are invisible and intangible. Descartes is surely right that we can't hope to define matter in terms of perceptibility, but it can hardly be an accident that matter occupies space in a perceptible way while fields "occupy" space imperceptibly. This indicates a significant difference in the way they relate to space. The suggestion about matter and space that I am about to make should not then carry over to fields and space.[7]

Explaining Impenetrability

One natural conception of exclusion is that bits of matter exert a kind of resistance to other bits of matter: when one object comes up to another object's place, that

---

[7] For a discussion of the ontology of fields, see Marc Lange, *An Introduction to the Philosophy of Physics: Locality, Fields, Energy, and Mass* (Basil Blackwell: Oxford, 2002). This book is, however, quite demanding for the novice.

object pushes it away, resisting its incursions. The object approaches, makes contact, and is the subject of a resisting force, which is matter itself. The movement of the approaching object is counteracted by a force from the given object, preventing it from occupying the same region of space. On this conception, exclusion results from an intrinsic property of the piece of matter doing the excluding—the resistance exerted by matter as such. It does not result from a property of the region of space occupied. Matter as such pushes other matter away, and space itself plays no role in the exclusion. What this absolute force is—and how it works—are not explained. It cannot be a repulsive force of the kind specified in physics because, as Locke says, no conceivable force could overcome it—the exclusion is necessary and logical, not causal. It is conceived as the resistant force of matter *qua* matter (and not, say, the resistant force owed to repulsive electric charge). Bodies differ from fields, then, in that fields exert no such absolute resistant force while bodies do, since fields are not spatially exclusive. Matter has a kind of irreducible and primitive power to interfere with other matter—to keep it fiercely at bay by its very nature as matter.

But there is another possible way to think about exclusion, which focuses on the space occupied not the occupying object itself. It is the space occupied that does the work of exclusion. Thus we have the idea that the occupying object modifies the space in some way—that occupation is an *operation on space*. It is then the space that is responsible for matter's impenetrability, not matter itself. (I mean here the space that is within the particles that make up matter—not the space that is between the particles that compose a macroscopic body: filled space not empty space.) The explanation of impenetrability is not that matter exerts a *sui generis* repelling force, but that the space occupied has changed its nature in some way as matter occupies it. It might be thought that the space comes to exert a repelling force, where before it did not, since space is by its nature receptive not exclusive; the space has been transformed into something that now resists movement. That is a logically possible view, though certainly a strange view: the locus of resistance is the occupied space not the occupying object. Of course, space has to be modified by matter for this to happen, which it does not need to under the first conception. The space within objects (including elementary particles) must have a different nature from the space between objects, according to the second view; but it can remain the same under occupation, according to the first view. I want to suggest a view of the second type, more as a hypothesis than a demonstrated truth, but the operation on space I have in mind is the rather drastic operation of *deletion*. That is, matter *annihilates* the region of physical space it occupies—puts it out of existence, nullifies it, makes it no more. Less drastic operations are conceivable: say, that matter removes a dimension or two from space, or makes space discontinuous and granular, or curves space, or. . . . But the hypothesis I want to consider is the most radical and simple: when a piece of matter moves into a region of space, it destroys that space completely. On this hypothesis, we have a quick explanation of impenetrability, namely that *there is no region of space there to be occupied*. The object that occupies it has put it out of existence, so *of course* no other object can occupy that place! It existed before the occupation, and will exist afterwards, but for the duration of the occupation it has been deleted from the spatial manifold. It is not that the region of space continues to exist and overlaps with the object from which it is numerically distinct; rather, there is no such overlap, because the region of space no longer exists. If you drop a marble into a bowl of water there is no water inside the marble; I am suggesting that there is similarly no space within (solid) material objects. But it is not that matter pushes space aside, as the marble pushes the water: that would make no sense—for where would the displaced region of space go? Regions of space, as Locke says (following Newton), cannot move, so they cannot be pushed aside by matter—space cannot travel through space! Instead, the space is extinguished for the duration. And a nonexistent space is a space that cannot be occupied. Thus, on this hypothesis, it isn't that matter exerts a repelling force on other matter; it is that the space is no longer *there* to be occupied. This makes the exclusion logical and necessary, not a matter of resistant forces. When an object moves and comes to rest it gradually eats up the space it finally occupies when stationary, making that space simply unavailable for further occupation. The object consumes space, putting it completely out of business. When we speak of an object occupying a given part of space, we are really speaking of the space that once existed now being taken up by a piece of matter. There simply is no space where matter is (though there is clearly extension). A possible world with no space in it would be a world in which matter could do no occupying; well, in the actual world there are pockets of spatial nothingness where no matter can go. Occupying space is erasing it. Of course, there is matter *in* space in the sense that there is space between pieces of matter; but there is no *internal* space.[8] That, at any rate, is the hypothesis, strange as it may sound. We must try to evaluate it on its merits.

Why is space receptive? Because there is no matter there, which is by nature exclusive. Why is matter exclusive? Because there is no space there, which is what makes receptiveness possible. This makes Lockean "solidity" an entirely different matter from hardness, since hardness consists in forces that hold constituent particles together in space, while solidity consists in the annihilation of space. It is not a matter of a force that cannot be overcome, but is written into the very structure of space. What reasons are there for accepting this theory and what reasons are there against it? I don't think any cast-iron proof can be given for the Deletion Theory, as I shall call it, though I think there are some intuitive considerations in its favor; but I also think the theory cannot be easily refuted. We should therefore give it a hearing and see where it leads us in terms of overall theory. In theoretical physics, strangeness is

---

[8] Of course, ordinary objects, like furniture and animals, do contain internal space, because the constituent particles are widely separated—but I am talking about solid continuous matter, such as composes the elementary particles. When I speak of objects in the text I am to be understood in this way.

no bar to truth: explanatory power and internal coherence rule. A first point, to be given no more weight than it warrants, is that we talk as if the Deletion Theory were true: we say that there is no space inside a region full of material stuff ("Is there any space left in the trunk?" "No, it's completely full"). My informal survey of naïve opinion has it that people do think that there is no space inside solid objects (they are aware that there is space between particles, but they don't think there is any space where matter is continuous). The folk tend to think it is as if matter pushes space aside, like air, but then they blanch at the notion that the space thus displaced goes anywhere. The notion of *full space* strikes naïve opinion as an oxymoron, once it is made clear that this implies an overlap of an existent region of space and a material object. Still, such opinions are inchoate and easily manipulated, so not much weight can be given to them.[9]

A second and weightier consideration is that if we allow the Overlap Theory then we are accepting that two distinct things can be spatio-temporally coincident—a region of space and a piece of matter. At any location where there is an object there is *also* the segment of space that the object now occupies, as a numerically distinct thing. The region of space was empty a moment ago, but now it is full—and it is just as much *in* that location as it was when vacant. We have a pair of overlapping entities, a chunk of space and a chunk of matter. But doesn't this violate the principle that there cannot be two things in the same place at the same time? It might be said that this principle is too restrictive anyway, because of the statue and piece of bronze kind of case. But that is a situation in which the statue is *constituted* by the piece of bronze, which makes sense of their coincidence; by contrast, the region of space is not constituted by the matter occupying it (nor vice versa), since destroying the matter doesn't destroy the space. So we have two entities that are not mutually constituting being said to be coincident. Of course, one of them is not itself a piece of matter (*pace* Descartes), but still they are distinct entities both numerically and constitutively—and yet they are said to overlap. This seems offensive to common sense and general principle, and it seems ad hoc to make an exception for the case of space itself. We avoid this if we adopt the Deletion Theory.

A third point is that there is a puzzle about exclusion if we accept the Overlap Theory. If the region of space that was once receptive still exists, as robustly as ever,

---

[9] Conversational implicature might be invoked to explain away this evidence: to say that there is space left in the trunk would be to conversationally imply that there is *empty* space there, which would be false—yet there might still be *filled* space in the trunk. It is hard to see what could settle the question, I admit. It strikes me as curious that common sense should be so up in the air on what seems like rather a straightforward question: is there space inside solid matter or not (physical space, not mathematical space)? Common sense is clear that there is *extension* wherever there is matter, but not whether this extension belongs both to matter and to the physical space occupied by that matter. The answer depends on how we intuitively think of space, especially how substantialist we tend to be about it—and we seem oddly uncommitted on the issue. Common sense has no worked out ontology of concrete space in relation to matter, it appears, just some rudimentary intuitions. It is, by contrast, quite clear that there can't be matter inside other matter—or two regions of overlapping space.

inside the occupying object, then why is it no longer receptive? Clearly it *is* not receptive, or else exclusion would not obtain. Its receptive powers have been completely used up, and yet it is still supposed to exist as a region of space with all its essential properties—so why does it not allow another object to co-habit with the occupying object? It might be replied that it is that object that is doing the excluding, in virtue of its intrinsic resisting power, so that the receptiveness of the internal space is being overruled. But this seems puzzling, since space is *essentially* receptive—so why can't it invite another object in, so to speak? So far as the space itself is concerned it is open for business, but the occupying matter insists on monopolizing it—the puzzle is why the matter gets to win this battle. By contrast, on the Deletion Theory, all is clear: there just *is* no space there to welcome another object. Once the first object has been admitted, the space is destroyed; it is just no longer available for occupation. On the Overlap Theory, though, the space marches on, as real as ever, and as intrinsically receptive as ever; yet it is puzzlingly confined in its hospitality to the object that *gets there first*. So far as the space itself is concerned, it could admit multiple objects, since its nature is to be receptive; but the occupying matter gets in its open-armed way. However, under the Deletion Theory the exclusion follows directly from the transformation matter effects with respect to space, namely its annihilation.

Here is a connected point: space and force fields do overlap, and neither is exclusive. A magnetic field, say, may be co-terminus with a region of space, and this region is as penetrable as any in space. Why? An obvious answer is that the existence of space is in no way compromised by the presence of a field, so that its usual receptive powers are preserved; thus material objects can intrude into this space. But if a similar overlap of existent and distinct entities characterized matter and space, wouldn't we expect a similar result, viz. receptivity towards other material objects? Yet we don't find this. The Deletion Theory steps in to explain way: fields are space preserving, but matter is not. It isn't that matter has some maximal repulsive force that fields do not have (indeed, wouldn't a field of *force* be the more likely thing to manifest such a resistant power?); it is that the very solidity of matter—its capacity to *take up* space—enables it to annihilate space by occupying it. This, at any rate, is an intriguing theory we should take seriously and explore. Matter takes a bite out of space, so to speak, so no other matter can get its teeth in (but it does allow the space to be born again when the matter moves elsewhere, so that a new piece of matter can take its consuming turn).

But what might be said against Deletion and for Overlap? Here is an obvious point: the space that an object occupies indisputably regains its existence when the object departs it. The Overlap Theory says that the region persists during its occupation, being present whether or not it overlaps an object. Space thus persists over time. But on the Deletion Theory it does not: it springs in and out of existence as matter moves around. More particularly, the piece of space that is extinguished by a lump of matter reasserts its being when the matter moves away. Space is thus continually popping into and out of existence as motion occurs. Curious, you might be forgiven for thinking. But is this an objection to Deletion? Only, I think, if we tacitly model the ontology of space on the ontology of matter. It would indeed be very odd to embrace a theory that had material objects popping into and out of existence—that is, the very same objects—since it is unclear what would make them count as numerically the *same* objects, as opposed to qualitatively similar ones (this is one of the puzzles of teletransportation). But space is not matter, and the resurrection of spaces cannot be compared to that of matter. For one thing, the criterion of identity is clear—identity of position in space (however this is to be understood)—and this allows us to declare that it is the same place that has now come back into existence once the object has fled. And there is just no reason to compare the process of putting a material object out of existence with that of putting a region of space out of existence; certainly, the latter is nothing like the dismantling of an object's material parts and their dispersal. So any sense of oddity here arguably derives from dubiously modeling the existence of space on the existence of matter. I really see no cogent reason to balk at this consequence of Deletion: space just is the kind of thing that can have the kind of intermittent existence in question. Selves, too, arguably have intermittent existence, because of the fact of dreamless sleep, so it is not as if there is no precedent for the interruption of spatial existence contemplated by Deletion.[10] In an entirely static universe, in which no occupied region were ever liberated from its possessor, I think we would be more inclined to accept the Deletion Theory, since we would have no experience of an area of space ever coming back into existence—we would never see an erased region magically retain its being as an object moves from it. But adding motion doesn't really change the picture, once we get over the idea that the existence conditions of space mirror those of material objects. So first ask yourself whether you like Deletion for a static universe, and then ask whether you really need to revise your view for a dynamic universe.

It might now be objected as follows: the essence of space is extension, but material objects are extended, so isn't the idea of internal space unavoidable? That is: pure space extends to an object's boundaries, then there is the extension of the object itself, and then space carries on extending outwards—so isn't there an overarching space that includes all three of these extended segments, with space continuing through the extended material object? How can there be no space within the extended object if the essence of space is extension and objects are extended? The answer to this is that extension is not the essence of space, if that means that every extended thing *is* a region of space: for matter has extension and is not (identical to) space. What has extension in the occupied region is a piece of matter; it does not

---

[10] In both cases we have individuating ties to persisting things that can anchor the continued existence of the intermittent entity: human bodies and brains, and objects in space. There are also things like intermittent wars, re-born countries, and revived clubs. The same entity can begin to exist again after a period of nonexistence. Continued existence does not logically require continuous existence.

follow that there is any extended *space* there. Matter has extension and so does space, but the extension of matter is not *automatically* the extension of space. The continuity of extension through matter is just the extension of *separate* things put end to end, so to speak. We can see this from the fact that it would not be possible to *identify* the occupying object with the space it occupies—because these entities have different properties (for instance, the space persists when the object is destroyed and regions of space cannot move while objects can).[11] First there is extended space, then some extended matter, and then more extended space: nothing compels us in these facts to postulate a space that *overlaps* with the extended matter. Everywhere in space there is extension, to be sure, but that extension can take two forms—of space or of matter. Thus it is possible for there be the same amount of extension in one region of the universe as another but a different amount of space, according as the two regions contain a greater or lesser amount of matter. The existence of extended matter is not *ipso facto* the existence of internal space. So there is no sound argument here against the Deletion Theory.

I therefore see no decisive argument against the view that there is no space internal to matter (of course, as I noted earlier, ordinary macroscopic objects are full of space—I am speaking of continuous units of matter). This raises the question of whether force fields are internal to matter: does the gravitational field extend within the basic units of matter, whatever these may be? Does gravity pervade impenetrable matter as well as penetrable space? I don't know of anything in physics that decides this question, and I see no present reason to insist that fields pervade matter in this way, as well as space. But I note that in principle fields *could* pervade both space and matter equally: maybe there is as much gravity and magnetism inside matter as outside it. What I don't see here is any reason to suppose that space must exist inside matter, on the ground that fields require space as their medium: for we might well suppose that fields can pervade *either* space or matter, so that there is no need to postulate on overlapping space for them to pervade—or we might simply deny that fields exist inside matter. Fields of force in physics are usually defined as regions of space within which a force operates. If the full story requires that such fields also encompass (i.e., pervade) matter, so be it—but there is nothing here to compel the introduction of space into matter. My own tendency here would be to say that matter (metaphysically) excludes fields as much as it does other matter and space itself (according to Deletion), but I don't want to be dogmatic on the point. But if we do lean this way, we can declare a principle of *strong* exclusion, as distinct from the weaker

[11] This is a Leibniz law argument against the claim that a piece of matter is identical to the region of space with which it overlaps. The object and the space have different properties, thus violating the indiscernability of identicals: the space could be occupied by another object, for example, though the object could not be occupied by another object (whatever that might mean); and the two have different conditions of destruction, as well as other property differences. The idea of overlap of numerically distinct entities is far more viable than the claim that the space just is identical to the object. It is thus impossible to argue that the space must exist where the object is simply because the space *is* that (extended) object. We cannot get the space to exist where the object is *trivially*.

kind that declares matter exclusive only of other *matter*. Matter excludes from its location other matter, fields of force, and space; nothing exists inside a piece of matter but itself. So far as I can see, nothing in physics or philosophy rules out such strong exclusion.[12]

Let me restate the Deletion Theory in the following way. Suppose God had the power to take chunks out of space: he touches them with his magic finger and a hole in space appears, varying in size. These gaps in space (which are not themselves spatial!) are dotted around and may be encountered by the chunks of matter God has also created. Then we can say two things about these gaps: (a) nothing material can get in there, since matter requires space through which to move, so that they would be utterly impenetrable; and (b) they would to all intents and purposes *be* a kind of matter, since *this is what matter is*. The gaps in space would be impenetrable and exclusive, but that is the definition of matter on the Lockean view. This seems odd to us because I introduced the gaps by saying that God created them by a subtraction from empty space, and we think of matter as an *addition* to empty space. But if the Lockean definition if right, then the gaps *would* qualify as matter no matter how created (and there is a distinct element of *per impossibile* in my supposition about their creation). In the next section, I shall further explore this conception of what matter ultimately is—a hole in space, to put it crudely.[13]

We can now say how extension and exclusion are related, and how the structure of space meshes with this. The extension of an object is simply the area of space throughout which it is impenetrable. It is not that objects are extensive *and* impenetrable, as if these were unrelated attributes; the extension of an object simply *is* the

---

[12] There is also the point from General Relativity that curvature of space is the medium of gravity, but it is hard to see how curved space could exist inside continuous matter. Curved space has causal powers, but how could it have those causal powers inside matter? It doesn't cause bodies to move gravitationally when it is sealed off in solid matter. What if the whole universe was composed of solid matter, so there is simply nowhere (no empty space) for gravity to influence the motion of bodies: could we still suppose that space has curvature inside this unbounded matter? If it does, it is quite idle. The right view is simply that there *is* no space inside matter, whether curved or not.

[13] A more nuanced position would be that there are two types of matter to consider: the negative kind consisting of pure holes in space, and a more positive kind in which something substantial is added to the pure hole—such as mass or electric charge. The point I make in the text is that even the first kind ("thin matter") would have the essential characteristics of material substance—so mere holes are sufficient for that. The conception of matter as punctures in space is not among the standard options considered by physicists and metaphysicians; in fact, I know of only two places in which the idea is mooted. Russell, in *The Analysis of Matter* (Dover Publications: New York, 1954), briefly raises the possibility only to dismiss it: "It would seem to follow that, if the electron is to have a definite position in space-time, it must be either a point or a hole. The former, however, is physically unsatisfactory, and the latter seems scarcely capable of an intelligible interpretation" (326). Paul Dirac, however, suggested that positrons be regarded as holes in space, where space is conceived as a sea of electrons: for a popular exposition, see George Gamov, *Mr. Tompkins in Paperback* (Cambridge University Press: Cambridge, 1965), chapter 14, "Holes in Nothing." But this is really not the same as the theory I am proposing, because Dirac takes the holes to occur in a physical manifold—the sea of electrons in space. Nor is Dirac trying to explain the impenetrability of matter by invoking the idea of holes in space (better: breaks or discontinuities).

extent of its impenetrability. There is no part of its extension within which it is not impenetrable (again, I am oversimplifying the empirical facts about macroscopic objects for expository convenience: as we know, most of their extent consists of empty space—the distances between the particles). Moreover, all parts of its extension are equally impenetrable, and maximally so. The reason for this correlation between extension and exclusion, according to the Deletion Theory, is that the extension of an object fixes the amount and location of the space it puts out of existence. Take any point within the extension of an object: that point makes non-existent the very region of physical space that corresponds to it, so that the entire set of such points determines the region of space thus annihilated. (Of course, there are still mathematical coordinates that identify these points—so there is still a "mathematical space.") The hole in (real non-mathematical) space that the object creates has precisely the shape of the object itself. This means that the area of extension is identical to the area of impenetrability, which is in turn identical to the area of space erasure. The last of these gives rise to the second of them, which then fixes the first: from spatial gap to exclusion to extension. The extension of an object corresponds to the size and shape of the cut it makes in space.

SPACE AND MATTER

There is a pleasing symmetry to the relationship between space and matter under Deletion. We think of (empty) space as consisting of gaps between material objects, areas of absence, as if objects were interesting islands in a dull sea of structure-less space. But we can equally think of objects as gaps in space, areas of absence, as if regions of space itself were the islands in a sea of matter. Space is an interruption to matter, but matter also is an interruption to space. Space marks the limits of matter, its boundaries, but matter also marks the limits of space, its boundaries. Imagine if the universe had a lot more matter in it and that macroscopic objects were materially continuous (not made up of particles in space): instead of being mostly space, like ours, this imaginary universe is mostly dense matter, with occasional areas of empty space. Given that matter and space don't overlap, this means that most of the extension of this universe is constituted by matter not space (this is how cosmologists think of the tiny early universe). Space would be the exception, the standout, with matter the routine condition. In such a universe space would seem special, a more positive presence than it seems to us in our mainly spatial universe; we would take it less for granted. Matter would strike us more as the absence of space, as opposed to striking us as the interesting exception to spatial monotony. And if the Deletion Theory is true, space really would be ontologically exceptional, a rarity. I think that then we would be quite inclined to view matter as the gap between spaces, the absence to their presence. Space would seem like a positive thing, not the pure negativity it now appears to us. Or imagine a universe consisting of an infinite amount of continuous matter, with no empty space at all (let's not ask whether this is really logically possible). According to the Deletion Theory, there is no (real, physical) space in this universe at all (but a lot of extension—and hence "mathematical space"). Now suppose that this universe suffers a big crunch whereby the matter begins to shrink inwards, thus exposing acres of empty space. Space comes to be where there was none before. Wouldn't this seem like the *creation* of something, the bringing into being of a positive reality? Instead of adding matter to space, we add space to matter. Then space would seem exceptional and remarkable and matter would seem the dull backdrop to an unfolding cosmological drama.

I am saying these far-out things in order to try to break a prejudice that views space as a mere absence, a nullity, with matter as the shining centerpiece of Creation. Let's try thinking of matter as the absence of space, as we now tend to think of space as the absence of matter. Just as we now speak of "empty space," meaning space with no matter in it (a vacuum), so we might speak of "empty matter," meaning matter with no space in it (dense or continuous matter). Actual material objects have lots of space within their boundaries, so they don't really *fill* the space they occupy; these objects are not empty matter at all, since lots of space pullulates within their boundaries. Only elementary particles are empty matter in the proper sense. We can evidently describe things this way, inverting our usual ontological invidiousness, and capture the facts just as well (assuming the truth of the Deletion Theory). As we think of space as holes between islands of matter, so we can think of matter as holes between islands of space: each can be viewed as a gap or break in the other. The continuity of matter is interrupted by the spaces hovering between objects, but equally the continuity of space is interrupted by those pesky lumps of occupying matter. Where matter leaves off space begins, but also where space leaves off matter begins. So instead of viewing matter as something added to space, try thinking of it as something subtracted from space. A gap in space is equivalent to the existence of a piece of matter, since both are constituted by impenetrability. Penetrability is the basic thing; where impenetrability exists we simply have the negation of penetrability—and this negation is the essence of matter. Think of it this way: God didn't first create space and then supply the world with a further act of creation in the form of matter, leaving his initial creation unmodified; he first created space and then decided to *take back* some of what he had created, putting punctures in the fabric of space—the result was matter, i.e., solidity in Locke's sense. (Of course, he had to do more to produce the specific properties of matter in our universe, such as mass and charge, but just to get matter in its most primitive form off the ground that act of subtractive divine hooliganism directed at space was enough.) To create impenetrability you just need to negate penetrability, and holes in space do that. But don't think of these "holes" as like spatial holes, i.e., cavities in matter; they are abstractly defined as discontinuities in space, breaks in its distribution. These are holes made purely of matter, in effect. To destroy (physical) space is to fill it, because of the impenetrability that results; and matter is defined as what fills space. What I am suggesting is that filling space can be thought of as a negative operation, just as we now think of the substitution of matter by space as a negative operation (as when the integrity of an object is smashed and its constituents scattered abroad). The creation of matter is essentially and fundamentally the destruction of space.[14]

Consider motion. According to Deletion, motion is the successive destruction and re-creation of regions of space (those that correspond to the extension of the projectile). The moving object, construed as solid, is a node of impenetrability; the medium of the movement is a manifold (for want of a better term) of receptivity. Matter is both what moves and what impedes movement; space is what is immoveable but allows movement. Space allows movement because it has the power to cease to be and come to be effortlessly. The receptivity of space—its power to permit unimpeded motion—consists in its existential flexibility: its ability to cease to exist when an object encroaches, and its ability to regain existence when the object has passed on its way. The property of space that explains the exclusivity of matter is the very property that makes matter able to move—its talent for temporary existence. Motion is like a rip that spreads through a garment, except that the rip is magically sewn up at its trailing edge: a traveling rip in space is a material object in motion. Movement requires both the presence and absence of space: presence, because there has to be space (empty space) to move through; absence, because the moving object itself must be constituted as impenetrable, and hence material, by the nonexistence of the space within it (according to Deletion).

Why do we tend to privilege matter over space, treating space as a mere absence? In doing so we make the existence conditions for space very easy to satisfy: we think of space as a mere vehicle for matter, so that wherever there is matter there must be space, internal as well as external. Thus we fall into the overlap conception, with continuous space virtually guaranteed by the presence of matter. If space is inherently an absence, then it can't take much to make it exist. It is as if God really had to do very little to create space, as the novelist has to do nothing creative to have a blank page in front of him. But space is *not* a form of nonbeing, a mere lack of anything better; it is a robust form of reality in its own right. God had to exert himself to get space off the ground, both as to quality and quantity.[15] Once we properly appreciate this, we see that the existence of space is not guaranteed by matter: there is nothing *automatic* about the existence of space inside matter—that is a substantive claim.

---

[14] Perhaps this is a good place to confess what may already be obvious, that my guiding methodological precept in this essay is not to shy away from the bizarre and paradoxical, so long as explanatory power and internal coherence are served. This is the very principle which has led to so much success in physics proper—don't be constrained by received ideas, try to imagine new ways of thinking. The radical idea may prove to be right, and at least it provokes the consideration of fresh possibilities. The physical world has never much respected human preconceptions, and seems to prefer the *outré*.

[15] The debate over whether space is Euclidian attests to its substantive character, because God had choices as to how he could construct space (and he turns out to be a non-Euclidian, by all accounts). Space has a specific geometrical character, so it cannot be a mere *tabula rasa*: it is not a featureless nothing. (The myth of the "bare particular" hovers in the background here.)

(As hitherto, I am speaking of real physical space here, not mathematical or coordinate space: the latter is not put out of existence by matter.) And the absence of space is the absence of more than an absence—it is the lack of something quite...substantial. I think there are two reasons for the habitual ontological downplaying of space, and thus two reasons for wrongly taking the existence of space for granted. The first reason is, ironically enough, the very centrality of space in our thought and imagination. We find it extremely difficult to form a conception of non-spatial reality (as Kant was famously aware), which is one reason why the idea of abstract existence perplexes us; we really cannot imagine a world without space.[16] Thus we tend to think that space comes with the territory, so to speak; it is just part of any reality we can imagine or conceive. It doesn't strike us as a substantive and optional ingredient of reality, but as an unavoidable background condition. Accordingly, we tend to assume that the interior of material objects must include real space—we spread space from outside objects to inside them because of its extreme ontological cheapness (as we suppose). Because space is the foundation of our thinking, we take it to be the foundation of reality, so that anything real must incorporate space; thus we assume that where there is matter there must always be space, through and through. If the Deletion Theory is right, though, matter excludes space—hard as this may be to conceive for our space-obsessed Kantian minds. (Of course, matter allows, perhaps requires, a *surrounding* space; the present question is whether it allows an *interior* space.[17])

The second reason we mistakenly regard space as a wispy absence, in comparison to solid matter, is simply the empirical appearance of the two. We can't see or touch (or smell or taste) space; we can see and touch (and smell and taste) matter. Space offends the empiricist in us, always a strident voice (and sometimes a quite hysterical one). Therefore we assume that space has minimal conditions of existence, as naturally etiolated—for it strikes our senses as vanishingly thin and insubstantial. We are apt to think that if our senses were confronted by genuine nothingness it wouldn't seem very different from the way space strikes us (whatever all this means). Thus we feel that there isn't much difference between space existing and space not existing, so that the Overlap Theory seems guaranteed as a matter of course: since space is nothing much anyway, viewed perceptually, it is *bound* to exist where matter does. Indeed, Overlap and Deletion come perilously close to metaphysical equivalence if space is a type of nonbeing to begin with.

---

[16] See P.F. Strawson's famous discussion of the no-space world in *Individuals* (Methuen: London, 1959).

[17] But what kind of space is far more controversial—how many dimensions, Euclidean or not, continuous or granular, relative or absolute. Also, maybe there could be a universe of solid matter with no empty space at all: the existence of a vacuum seems metaphysically contingent. Extension is arguably essential to matter, but surrounding empty space seems accidental. But it is difficult to be certain, because we know so little about matter (see later essays in this volume). Is it really just a matter of chance that matter doesn't occupy *every* region of space? (Mass would presumably be extremely high if that were so, and hence gravity.)

How can you reduce what is nothing to less than nothing? How can you put an already non-existent thing out of existence? But I take it that this is all horribly confused: for space to exist is *not* for there be a type of nothingness pervading the universe, as if there were no real difference between space existing and space not existing. That difference is as large and momentous as any difference between existence and nonexistence could be. If space did not exist, the universe would be *massively* different from the way it is now. Whether space is the kind of existent our senses respond to as they naturally do to matter is beside the point; metaphysically, space is a robust existent entity, a definite component of reality. (Space, I feel like shouting, *IS*.) Thus it is a substantive question whether the conditions of existence for this entity are met in the interior of objects, and I have tried to suggest that it is not clear that they are. I think we should regard matter and space as ontologically on a par: both have substantive existence conditions, with definite natures, and neither should be regarded as the mere absence of the other. Then we can be open to the idea that there is an important symmetry between matter and space, with space as the interruption of matter and matter as the interruption of space. Each is a break in the continuity of the other. Each is a type of gap from the point of view of the other. We can describe the universe in either way and do justice to the facts that need to be captured.[18]

MIND, MATTER, AND SPACE

Descartes had three purposes in his definitions of mind and matter: to arrive at an accurate definition of matter, to arrive at an accurate definition of mind, and *to make sure that these had contrasting definitions*. His conflation of matter and space, much derided by Locke and others (rightly, in my view), should be seen in the light of that third purpose. For, from the point of view of distinguishing mind from the rest of the natural world, his definition of matter as extension works well enough; it is of no serious import if matter and space get conflated, so long as mind emerges as categorically distinct. Indeed, contrasting mind with space is just as important to Descartes as contrasting mind with matter, since mind is construed as essentially lacking in extension, and neither space nor matter are characterized by thought. Locke, on the other hand, does not fashion his definition of matter with that goal in mind, and so he is far more interested in correcting Descartes' conflation of matter and space. And it is a real question how Locke's impenetrability criterion fares when it comes to defining mind. I propose to make some brief remarks about this in the present section.

---

[18] This is a case of high-level empirical equivalence, where the data don't decide between the two modes of description. We can describe things either way and not conflict with observation. Yet one theory may have an explanatory advantage over the other—as the Deletion Theory can explain impenetrability, while the Overlap Theory cannot. It is a mark of how little we know of physical reality that such underdetermination of theory by evidence is possible.

Descartes says something interesting about the question in his reply to Henry More[19]: "We can understand easily that it is possible for the human mind, God, and many angels to be in one and the same place simultaneously" (168–9). He takes this to show that such entities do not have extension; whether it does show that is debatable, but what it does clearly show, if correct, is that such entities are not exclusive or impenetrable. And this would establish, presumably to Locke's satisfaction, that the entities in question are not material. Now, I don't propose to engage these historical figures in a dispute about God and angels, but it is notable that Descartes does not say that many human minds could be spatio-temporally coincident. He seems to avoid this question—and that may be because of its awkwardness for him, and indeed for Locke. For it is by no means obvious that many human minds, or selves, *could* be thus coincident. Of course, it would be trivially assured that this could not be so if we assume a general materialism about the mind—that is, if we assume that minds are material in the sense that they are impenetrable. But what is more interesting is that, even under a dualistic metaphysics, it is not clear that minds can share their location. So it is not clear that they are really an exception to impenetrability. The problem is not that materialism will follow trivially from the fact of their exclusiveness; the problem is that the definition of matter is in danger of being too broad to capture what is distinctive of matter. Since we don't want materialism to follow trivially from the very definition of matter, combined with the exclusive nature of minds, we need to be careful that it doesn't include the mind inadvertently. We surely don't want to find ourselves saying that the mind is a material thing just because your mind and mine cannot occupy the same region of space at the same time. Nor do we want the definition of matter to include mind so easily.

I think the right response here is to question the idea that the mind occupies space at all or has genuine spatial location, so that the question of overlap becomes moot. What this means is that minds do not have extension, at least as a matter of definition, whether or not they exhibit any analogue of material impenetrability. So we can't say they are exclusive over the range of their extension, because, unlike matter, they lack extension. Maybe it is true to say that there cannot be two or more minds operating at a specific locale simultaneously, but this is not to say that there is impenetrability over a specific spatial extension—which is what the proper definition of matter asserts. So we need to refine the definition from simple exclusiveness to exclusiveness *over a specific extension*—then we can avoid including minds as material for trivial definitional reasons. We make it possible to be a dualist while accepting that minds cannot interpenetrate. Of course, what such interpenetration might even be is extremely unclear, so it is unclear what we are *saying* when we declare minds exclusive; yet there seems to be some intuitive content to the notion. In any case, the exclusivity of minds with respect to other minds does not by itself commit us to a form of materialism or threaten the viability of a Lockean definition of matter.

[19] Work cited in note 2.

It is a further question whether minds or selves are exclusive of material bodies, granted they are exclusive of other minds or selves. This seems less clear than the case of other minds, for two reasons. First, it does not seem obviously wrong to say that I am where my body is, so that the self (which is not identical to the body) shares the body's location. True, it sounds funnier to say that the self is co-terminus with the body, sharing its extension. But we don't immediately recoil at the thought of self and body occupying precisely the same volume, so the self seems receptive at least to the material object that constitutes a person's body. Secondly, there seems to be some sense in which objects can enter the mind in acts of intentionality: the mind incorporates external objects into its content, so that there seems some overlap between the mind and material things. Again, this is murky and controversial, but there does appear to be something to the idea that the mind can be penetrated by objects. It is not that such objects are kept beyond the boundaries of the mind by some sort of barrier of impenetrability. If this is right, then the mind lacks the kind of absolute impenetrability that characterizes matter, and hence falls outside of the definition of matter I have been working with. This seems the right result, on balance.

There is a kind of abstract analogy between mind and fields of force, in that neither is exclusive of material objects, and both bear an obscure relationship to space. Minds and fields relate to space differently from the way regular matter does, since they don't take up space yet are roughly locatable within a spatial region. Neither can be said to annihilate space, since neither precludes the presence of objects within their boundaries. But this rough analogy is no more than that: it would be wrong to press it too far, and absurd to suggest that mind is really a kind of field like an electrical field. Yet it is intelligible that some people should find in the notion of field a physical counterpart to consciousness—because consciousness has more in common, formally, with fields than with lumps of matter.[20]

I believe we should work with a fourfold ontological taxonomy: space, matter, fields, and mind. The first three items here exhibit important ontological differences, especially matter and space. It is crude and misleading to speak dichotomously of "the mental" and "the physical," as if all but the mental belonged to a single category. Matter is fundamentally different from space and to classify both as "physical" simply blurs distinctions and serves no wider purpose; the notion of "the physical" begins to seem artificial and tendentious, not a reflection of deep ontological distinctions. The basic distinctions here revolve around notions of receptiveness and exclusivity; and space, matter, fields, and mind should be classified according to this basic opposition. The whole notion of "the physical" is a kind of Cartesian holdover,

---

[20] There is a lot of New Age nonsense about consciousness being a field that surrounds the body, which field can act directly on things outside. This is scientifically silly, no doubt, but it is naturally suggested by the concepts. Galen Strawson interestingly plays with the idea of consciousness as a force or field in *Real Materialism* (Oxford University Press: Oxford, 2008), 43–44.

reflecting Descartes' misguided attempt to assimilate matter and space.[21] Locke's critique of this is a step in the right direction, suggesting as it does that we need more than a two-way distinction. Once the radical difference between matter and space is appreciated, the attempt to divide the world into "the physical" and "the mental" looks deeply wrongheaded.

Finally, the conceptual apparatus developed here has a bearing on doctrines such as panpsychism, neutral monism, and idealism. These doctrines attempt to construe the nature of matter along mentalistic lines, taking objects to have a kind of mental essence. Russell's doctrine that the intrinsic properties of matter are mental while its structural properties are described by physics is a case in point.[22] The trouble with these views, from the present perspective, is that it is not clear that they can fully accommodate the impenetrability-over-extension that is so crucial to the nature of matter. To the extent that mental phenomena fail to fit the model of exclusion within a determinate extension, they are unable to deliver the real essence of matter. Matter is of a radically different ontological category from mind, because of its special relation to space, so doctrines that edge matter in the direction of mind will be unable to capture the essence of matter. Berkeleyan idealism, for instance, with its ontology of ideas and spirits, seems unable to find a place for the kind of impenetrability so crucial to material bodies.[23]

PHYSICS AND MATTER

What is the bearing of the present theory of the nature of matter and space on the science of matter and space, viz. physics? Physics tells us the laws and constituents of matter, but it does not tell us how to *define* matter—that is a job for philosophy (of course, physicists may also be philosophers). (Much the same might be said of psychology and the definition of mind.) In some broad sense, we are trying, as philosophers, to elucidate the *concept* of matter—to analyze it. Whatever answer we give to that conceptual question can then be plugged into physics itself, and it may well determine the content of our physical theories. We thus arrive at an *interpretation* of physics, driven by largely philosophical concerns. Let me illustrate this procedure, as

---

[21] This point has long been urged by Chomsky: see the entry by him in *A Companion to the Philosophy of Mind* (Basil Blackwell: Oxford, 1994), ed. Samuel Guttenplan. It is striking how self-described "physicalists" just assume that space is physical—without adopting Descartes' view of space as a material plenum.

[22] See Russell, *The Analysis of Matter*, cited in note 13, esp. chapter XXXVII, "Physics and Neutral Monism." He says: "As to what the events are that compose the physical world, they are, in the first place, percepts, and then whatever can be inferred from percepts by the methods considered in Part II" (386).

[23] Why should *ideas* be impenetrable by other ideas? Can ideas have an extension over which their impenetrability holds? Are spirits (finite and infinite) impenetrable over some specific extension? Berkeley holds that there is no such thing as matter, so nothing is truly impenetrable: but it is a datum that ordinary bodies are impenetrable over their extension. So idealism is false.

I understand it, by examining some basic concepts and principles of physics with the present philosophical theory in mind; I think we shall see that there are some interesting consequences.

(i) I have already talked a bit about motion, but now I want to spell out how motion is to be interpreted under the present theory. The movement of matter through space is probably the most fundamental process in our universe—perhaps any universe. Physics seeks to state the laws that govern motion, mathematically if possible, and it has arrived at a successful formulation of these laws. I am interested, however, in the *analysis* of motion—the necessary and sufficient conditions of its possibility. I want to know what motion *is*. In order to answer that question we need to grasp the nature of matter and space (as well as time), since these are the twin pillars of the existence of motion: motion simply is the change of position of *matter* through *space* over *time*. According to the present theory, the essence of space is receptivity or penetrability and the essence of matter is exclusivity or impenetrability. Matter both moves and hinders motion; space both does not move and permits motion. Matter might be defined as a moving impediment to motion; space can be defined as an unmoving facilitator of motion. Space is homogeneously receptive: no area of it is more receptive to motion than any other area (unoccupied space, that is). The receptivity of space is absolute and constant: space is maximally receptive and it is so in every part of it. Nothing about space itself can ever retard motion (as opposed to the forces or matter that might exist within it). Matter, by contrast, is absolutely and constantly impenetrable: matter is as impenetrable as can be and all matter is so. These are, if you like, metaphysical truths about space and matter—inherent necessities. Movement then is essentially the penetration of the essentially penetrable by the essentially impenetrable: that is, it is the penetration of *successive* regions of penetrability by impenetrability over time. But since the moving object is impenetrable, and essentially so, moving objects can collide: pockets of impenetrability impede the free motion of other such pockets. Collision is just the meeting of one impenetrable by another impenetrable in a vast sea of penetrability. An object blocks off a region of space from another, and a collision occurs. The transmission of motion from one object to another by means of collision results from the fact of impenetrability: if pieces of matter were not impenetrable, then collision could not occur and the transmission of motion could not occur. Motion cannot be transmitted to a penetrable thing like space; this is possible only in a universe where there are pockets of impenetrability. Only where there is spatial exclusion can there be collision and motion transfer—or else one thing simply passes through another, as matter passes unimpeded through space. The reason space is immoveable is that it is perfectly penetrable; the reason one object can transmit motion to another through collision is that matter is impenetrable. Penetrability and impenetrability are metaphysically basic in the analysis of motion.

But there is a further dimension to motion, arising from the Deletion Theory. Motion is possible only because space is existentially flexible. Since material occupation is equivalent to spatial destruction, movement is the successive destruction and then re-creation of regions of space. The status of the moving object as impenetrable depends upon the deleting operation matter performs on space, but since movement is the successive occupation of contiguous regions of space, this deletion must occur at every point in an object's trajectory—and the subsequent re-creation of that region of space must also occur. Temporary annihilation is the basic process. The inert storage of objects in (surrounding) space involves deleting a region of space, but the change of location of objects depends upon performing the operation repeatedly—while not *using up* the space that is left behind. A universe without motion, with space limited to its storage function, would fail to exploit one of the key facts about space—its existential flexibility. If God had made space and matter but not motion, he would have wasted a key power of space: storage requires the ability of space to be destroyed, but motion requires that space be capable of resurrection (and we know God has a keen interest in resurrection). Space simply cannot stay dead once matter has departed from it—its very nature is to exist when unoccupied. Indeed, it is hard to imagine killing a region of space *without* occupying it—what would one set out to do, starve it to death? Motion is guilt-free destruction of bits of the cosmos, since space is no sooner killed than it becomes alive again. We might even say that space is *designed* for motion, as well as storage, so that a static universe would be failing to live up to its purpose (or at least potential). The nature of space *predicts* the possibility of motion. Space actively encourages motion.[24]

(ii) Physics is rife with conservation principles: energy of all kinds is conserved, and so is matter (since it is regarded as interchangeable with energy). Whether space is conserved is more controversial: those who believe that space itself expands with the expansion of matter brought about by the big bang presumably deny it—though one can imagine formulations of big bang cosmology that preserve the common-sense idea of spatial conservation. At any rate, it is assumed that motion itself does not change the quantity of space in the universe. But the Deletion Theory, at first sight, seems to violate this assumption, because space is destroyed by being occupied. However, it is easy to see that this is not so, since for every piece of space destroyed another regains its existence: when an object moves into a new place it destroys that place, but it allows its old location to spring back into being. If the quantity of matter in the universe is constant, then no amount of motion will lead to a decrease in the amount of space, under Deletion, since all motion is destruction

---

[24] A meta-philosophical note: I have here proceeded by restating the familiar facts of motion in terms of my analysis of space, matter, and material occupation—no new facts have been adduced. I take this to be the old method of conceptual analysis: paraphrase, expansion, and elucidation. Such analysis is essential to philosophy, in my book.

and resurrection. It is true that there is *less* space in the universe under the Deletion Theory than the Overlap Theory, but the Deletion Theory does not lead to the consequence that the amount of space can vary over time. It is also true that if the quantity of matter were to increase then the quantity of space would decrease, which is not so under the Overlap Theory, but the overall quantity of matter does not vary because of the conservation of matter (though it can become denser or more dispersed). Would it be worrying if space did vary in this way? No more so, I think, than if matter also varied; so this is not a consequence that should make us rethink the theory. Interestingly, we can also invert this relation between the conservation of matter and space: *given* that space is constant, matter must also be, because matter can only vary in its quantity if space does. If matter did vary, then space would, but space can't, so neither can matter. We can't give this argument under the Overlap Theory, since an increase in matter involves no decrease in space. But we can deduce from the premise that space cannot decrease the conclusion that matter cannot increase (similarly for increase of space and decrease of matter). The two principles of conservation are tied together under Deletion, as they are not under Overlap. I think this favors the Deletion Theory to some degree, since theoretical integration is always a good thing. We also have it that the amount of solidity in the universe is conserved: impenetrability cannot increase or decrease, and neither can penetrability. The "solidity value" of the universe is a constant.[25]

(iii) I said that every part of space is equally penetrable. Does this have any bearing on the question of the infinity or boundlessness of space?[26] Well, if space had a boundary, a perimeter beyond which it was impossible to go, wouldn't this mean that it was *not* penetrable at every point? If an object in motion reached this boundary, it could go no further, since there is no motion without space: but wouldn't that mean that the outer shell of space *impeded* motion? If it did then a part of space would not be penetrable—the boundary itself. But every part of space is penetrable, *so* there cannot be such a boundary. We are familiar with the argument that no matter how far an object travels it could always travel further (hence space is infinite or boundless); I am suggesting that a bounded space would violate the total penetrability of space—a rather different argument. The way to resist this argument is to maintain that the boundary of space is not a *part* of it, so that a limit to motion there is not a limit imposed by anything intrinsic to space. But it seems hard to maintain that position, given that the resistant edge of space would *have* to be integral to it: when the moving object stopped traveling, by dint of having used up all penetrable space, this would have to be a result of something about space itself, so space would

---

[25] I mean the total quantity of solidity, not the individual solid things. The amount of solidity stays constant, not which things are solid. It is much the same with energy: the total is conserved, not the individual instances.

[26] The disjunction here is simply to allow for a finite but unbounded space—such as physics and mathematics often introduce. Total penetrability of space strictly requires unboundedness, not infinity—though I myself tend to favor infinite unboundedness, not the finite kind.

have to be functioning as if it were matter, i.e., exclusively. To be halted by space itself is contrary to the very essence of space; therefore there cannot be such halting—which is to say that space is infinite or boundless. However, these questions are very obscure, and it is hard to know how much confidence to repose in such arguments, attractive as they may be.

(iv) Quantum theory is another obscure topic, and I will make just a few brief and unsatisfactory remarks about it. It may be said that the whole idea of matter as occupying space is brought into question by quantum theory: particles of matter have no definite location, but only a range of probabilities of position within some region (the "uncertainty principle"). We simply cannot say that a particle is in one place rather than another, so there is no definite place from which it excludes other matter. To this I reply that, even accepting this kind of radical indeterminacy, we can still say that there is a principle of exclusion at work, since particles will exclude each other from whatever *indefinite* location they may have. No two particles can occupy the same indefinite region; or again, whatever type of fuzzy location particles may have, it cannot be that two or more of them have it simultaneously. Particles cannot be exactly coincident while being inexactly located. Here it may be said that I am clinging to a conception of the real nature of matter that quantum theory has made obsolete, namely the idea of classic Newtonian particles that compete for location. According to quantum theory, so-called matter is just a pattern of waves, a collection of forces or energies, with no solid material substrate. If so, at the most basic level there really *isn't* any matter, not in the antiquated sense I have been assuming; it's really all forces down there, fields of free-floating energy.

Now there is a lot to be said about such an interpretation of quantum theory, but I don't propose to question it here. Let us concede for the moment that forces and energies lie at the root of things, with solid classic particles not part of the story. This immediately implies, given what I said earlier about fields, that the basic level of physical reality does not exhibit strong metaphysical exclusion, since fields do not compete with other fields for spatial occupation. So, according to this version of quantum theory, there is no impenetrability at the foundation of the physical world—coexistence in space and time is not theoretically ruled out. However, at the macroscopic level (and this means at least from molecules up) there *is* clearly exclusion; certainly, the lack of it at quantum levels does not entail its lack at higher levels. We simply do not observe the interpenetration of large-scale objects, and this cannot be some kind of giant cosmic coincidence. Nor, in fact, is there any evidence of electrons and protons piling up and overlapping in one place. (I discuss this kind of question further in the postscript to this paper.) What I would say in reply to this line of argument is that solidity (in Locke's sense) might be an *emergent property* of matter: it is not to be found at the basic level (at least so we are conceding for the sake of argument), but it is found at higher levels—so it is a new property not predictable from the lower levels.[27] And this emergent property is what I (and Descartes and Locke) am trying to define. We cannot assume that the properties of the quantum level are the *only* properties of the physical world, and if solidity is not found there it must come in at another point (compare consciousness). As with all kinds of emergence, there is an element of mystery about this, but the facts seem to demand it: the matter we observe is impenetrable, and necessarily so—you cannot have two material objects, such as two cats, in the same place at the same time. (It is a different question whether we can in principle crush all the objects of the universe into a tiny area—a possibility I do not dispute. But that does not entail material coincidence in the sense in which I deny it.) The classical notion of solid matter *does* apply at the macroscopic level, no matter what may be true of the quantum level. It would be a mistake, then, to think that quantum theory makes the enquiries here undertaken obsolete—a philosophical mistake. In addition, of course, the interpretation of quantum theory I am conceding for the sake of argument is highly questionable; and I myself am not at all convinced that the classical notion of matter as extended impenetrability is seriously brought into question by the bare facts of quantum theory. My point in this section, though, is just that we have some options and wiggle room in reconciling quantum theory with the theory proposed here (or some modified version of it: see the postscript).

(v) How does the conception of matter I am working with relate to the physicist's notion of mass? What about the idea that matter can be *defined* in terms of mass? The concept of mass is supposed to capture the idea of the quantity of matter in a given object, so that we can say what it is for one object to have *more* mass than another. Mass, as a physicist understands it, is basically weight corrected for gravity, where this is explained in terms of inertia: mass is measured by the amount of acceleration produced in an object by a constant force. This is quite intuitive: the more massive an object is the harder it is to move it with a constant force; the heavier, the slower, to put it at its simplest. The point of the notion is to permit comparisons of mass, but we could adapt it to provide a potential definition of matter: x is matter if and only if x has mass. There are, however, a number of problems with this as a definition. First, we cannot explain mass by adverting to the notion of quantity of

---

[27] It is often said that elementary particles lack Cartesian extension, being essentially field-like. Nevertheless, they compose bodies that have such extension. So macro-extension is a kind of emergent property, resulting from combination. Some such story could be told about solidity: it is not present in the particles considered singly, but arises when they are combined. Another possible response to radical ontological interpretations of quantum mechanics is that physics has shown that there *is* no matter in the world, since matter is by definition impenetrable and there is nothing impenetrable out there. Objects are simply not made of matter, because they lack the exclusivity definitive of matter, according to quantum theory. Then, it is not that the impenetrability analysis is mistaken: it is just that there turns out not to be anything satisfying the analysis. There may be matter proper in other possible worlds, but in the actual world there is none—just energy or force fields. I don't find this an attractive view, but it is worth noting that it is logically available and sometimes endorsed—and does not strictly undermine the Lockean account of what matter proper is.

matter, or else we run in a circle (x is matter if and only if x contains a quantity of matter). Suppose, then, that we adopt the physicist's (operational) definition in terms of inertia, so that having mass is showing some (non-zero) inertia under a force. Then we have an initial problem about excluding space, since no force can move space and yet it is not material (does space have infinite inertia?). We clearly can't stipulate that the force must be applied to a piece of *matter*. Also, can there be mass-less particles that exhibit both extension and impenetrability? That doesn't seem conceptually ruled out, and one hears that such things may actually exist (e.g., neutrinos). Furthermore, the definition implicitly appeals to solidity, since no object can *resist* a force unless it is solid (this is how we would naturally rule out the space counterexample). We apply a force to an object and observe how it accelerates, but it will only do so if it is solid, i.e., impenetrable—or else the force will go right through it encountering no resistance. (Strictly, the degree of resistance needs to be non-zero: it does not have to be absolute—though, by strong exclusion, it will be. But we can consider worlds in which impenetrability comes in degrees: see the postscript.) The putative definition also presupposes the concept of motion, which depends upon the idea of an object in motion, i.e., matter. For philosophical purposes, then, the concept of mass is not what we need, however important it is for the purpose of the physicist—which is to provide a *measure* of (relative) quantity of matter. The amount of matter in an object certainly determines its propensity to move, but we cannot really *define* matter as what has a propensity to move (or not move) under the application of a force. Matter consists fundamentally in impenetrability, where different impenetrable things may have different masses. It is better to define mass as the amount of impenetrable substance in an object, if we want a measure of quantity of matter. This will then deliver a numerical result for inertia, since the amount of such substance affects its inertia. In any case, the physicist's notion of mass is not what we need to capture the concept of matter.[28]

(vi) How are we to understand gravity within the current framework? Gravity causes matter to condense, among other things, to cluster into larger and denser lumps; thus the formation of the galaxies from dispersed and formless gases. Gravity pulls bits of matter closer and closer together by its attractive force. But there is a limit to this collapse—matter can only get so dense. And that limit is imposed by solidity: given that the basic particles are impenetrable, there is a logical limit to how densely packed they can become—they cannot cluster any more densely than contiguity. They cannot approach each other's position any more closely than their extension permits; there cannot be overlap. No matter how powerful and overwhelming the force of gravity becomes, it cannot collapse matter any further than exclusion allows. (This is consistent with collapse into a finite point, but not with collapse into

---

[28] Just as defining matter by means of the possession of electric charge or particulate structure would not be satisfactory—any more than defining it by perceptibility is. These are extrinsic contingent properties of matter, not what constitutes its essence. In chapter 2 I consider the question further.

a mathematical point, given that particles have finite size.) So impenetrable things are the source of gravity, since they have mass, but they also set a limit to its power: even the densest black hole cannot contain spatio-temporally coincident pieces of matter. In a universe in which the source of gravity consisted of penetrable units (*per impossibile*), there could be complete overlap of units with no limit to gravity's condensing power, but any universe with *matter* in it can permit just so much gravitational collapse. We are not going to discover super-black holes in which matter is so compacted by gravity as not to be exclusive at all.[29] It is, however, possible in principle that gravity (or some other force) might *transform* matter into something else—say, thermal energy—and that this new stuff might be compactable indefinitely: but that would not be matter as we know it, perhaps being more akin to a force field. If matter came from the initial singularity, but was not itself present at the very onset of the big bang, then the origin of all matter might not have exhibited exclusion—being perhaps just a superhot quantity of energy. (Or we might consider the state of the universe *before* the big bang: the "stuff" then might not have been impenetrable, and not have been matter, though it gave rise to matter as the universe evolved.) But the matter that emerged from the big bang had the (emergent!) property of impenetrability, and so it posed a limit to how much it could be compacted. Matter can indeed become extremely dense, with little or no space separating its constituent particles, but it retains its essential exclusive power even in these extreme conditions.

According to Deletion, matter creates "holes" in space (in space-time if we are being *au courant*). The more matter there is, the bigger the hole created. The sun creates a much bigger rip in the spatial fabric than the earth. But the sun also exerts much greater gravitational force than the earth. So the bigger the rip is, the greater the gravity. We can thus say that the source of gravity *is* gaps in space: objects are attracted to these gaps, with greater force the greater the gap is. It is hard to avoid the metaphor of being sucked into a hole, of falling into a gap. If we ask *why* matter exerts gravity, we can reply that it does so because it creates holes in space into which other matter gets sucked. So we get a picture of gravitational fields as something like Cartesian vortices, with the matter at their center acting as a sinkhole. The more matter gets sucked in, the bigger the hole becomes, and the more it sucks other matter in; which makes the hole bigger. Gravity is in fact a kind of sucking action, because attraction is a pulling force; and the pulling is towards a kind of absence. Whether this is more than a metaphor is hard to say, but certainly the idea

---

[29] Sound metaphysics can have empirical consequences: we shall never discover super black holes with matter so condensed as literally to overlap with itself. Similarly, we shall never discover a surface that is simultaneously red and green all over, or a person with no brain and body, or empty space that resists occupation, and so on. There are a priori limits on how the world can be discovered to be. (I am well aware of how difficult it is for some philosophers to accept this—the all-out hardheaded empiricists among us.)

of matter as spatial deletion leads to a picture of gravity as attraction towards a chasm (a chasm *in* space not *of* space). Gravity results from the absence of space, which is the same thing as the presence of matter: so matter moving towards a gravitational body is rather like liquid drawn into a sinkhole. According to General Relativity Theory, matter bends space and gravitational motion results; according to the Deletion Theory, matter causes a break in space and gravitational motion is in the direction of the break: in both theories matter operates on space to change its nature (its geometry and causal powers). Space itself plays a role in the workings of gravity, for both theories, not merely the masses of matter dotted around it—the curvature of space (space-time) and the discontinuity of space.[30] Here we see space as an active agent, not merely a passive receptacle: it is not ontologically indifferent to the presence of matter, serenely retaining its permanent structure, but actively responsive to the presence and motion of matter. Space is an alterable, active medium, not a static backdrop—an actor on the cosmic stage, not merely the unchanging stage itself. Discontinuities of space are the source of matter's impenetrability, and this impenetrability is what makes the world go round—that is, constitutes matter as what it essentially is. Combining General Relativity with Deletion, we get a picture of gravitational motion as consisting of rips in space moving according to the curvature of space: space is where the fundamental action is, with matter along for the ride.

I said just now that the bigger the rip, the stronger the gravitational force: but is that quite right? First, let me introduce the idea of the *proper extension* of an object. Objects as we find them are composed of particles in interior space, with some objects more dense than others. The holes in space created by objects do not extend over their ordinary perceived extension; only the constituent particles create holes—we might better say, *nicks*—in space. A picture of the spatial hole made by a golf ball, say, would look like a scattershot, with more internal space than holes in that region. The more widely spaced the particles are, the less dense the object, and the less damage it does to the fabric of space. Let us then say that the *proper* extension of an object is the extension it would have if all its particles were made contiguous (or placed at some constant distance). So a more voluminous object could easily have a smaller proper extension than a less voluminous object, if its particles were less densely arranged. We could then speak of the quantity of matter contained in the proper extension of an object, and this would be a way to characterize the mass of the object (Newton understood mass as volume times density). If two objects have the same proper extension, then surely they carve the same size hole in space, since they contain the same amount of solid matter. So is it correct to say that gravity is a

---

[30] In fact, the discontinuity causes the curving, if we define matter in terms of Deletion: the larger the hole cut the more the curvature. That's interesting: it is as if the rip in space induced by occupying matter ripples out through space, deforming it. The puncturing brings about the bending.

function of the size of hole made by the proper extension of an object (which is equivalent to its mass)? The problem with this formulation is that it makes no allowance for the potentially varying mass of the individual particles. Suppose one object has particles with generally greater mass than another object, in the sense that it is harder to accelerate them under a constant force. That means that the two objects could have the same proper extension (the same volume of contiguous matter) and yet have different masses (ease of acceleration). In other words, they would cut the same size hole in space but one has more mass than the other. Given that gravity is held to be a function of the mass of an object in the inertial sense, it follows that gravitational force can vary though the hole incised in space is the same. Therefore, gravity is not a function merely of the size of the hole, but of the mass of the particles that make the hole. How should we respond to this? Here is one possible response: the holes may also vary in how *deep* they are. That is, we suppose that more massive particles dig deeper holes, so that the holes in space are actually bigger than is fixed merely by proper extension. Now this notion of depth is *not* the same as ordinary depth, since the holes cut in space by three-dimensional objects are also three-dimensional. What we have to postulate is an *extra dimension* in space to accommodate the effects of mass at the particle level: the more massive the particle the more it intrudes destructively on the fourth dimension (or nth dimension). I mention this not as an empirically justified conjecture, though postulating extra spatial dimensions is not unheard of in theoretical physics, but as a possible way to preserve the idea of gravity as a function of the geometrical magnitude of the spatial gap cut by an object—given that gravity is a function of inertial mass and that particles can differ in this respect though they have the same extension. It is an intriguing idea, though certainly a highly speculative one: particles of equal volume in three dimensions, but with different masses, have different volumes with respect to a further dimension of space, so that mass and volume stay tightly connected. We have got (somewhat!) used to entertaining the idea of extra dimensions to space in String Theory; but now there is a reason that might prompt a theorist to add an extra dimension in order to get a nice neat theory of space, matter, and gravity. It requires that we tinker a little with the idea of deletion, giving it an extra dimension to play with, with the notion of volume extended into an extra dimension. Cosmology, however, is not in general without such extravagances, and physics has taught us that we should always keep an open mind where matter is concerned. Matter is full of surprises.

CONCLUSION

It is curious to me that the question of the reality of space internal to continuous matter does not concern physicists more; indeed, so far as I know they don't even raise the question, let alone take a stand on it. Perhaps the operationalism and verificationism and mathematization that have come to characterize the field put such questions out of professional bounds.[31] Nevertheless, answering the question one way or the other has quite substantive consequences for the way we think about the physical world, and opens up interesting theoretical possibilities. Perhaps the question is just too "philosophical" for hardheaded scientists, but it seems to me genuine and difficult to answer. I think, at least, that the Deletion Theory should be taken seriously and considered on its merits, however strange and counter-intuitive it can seem to be (as well as common sense in some moods). It is true that it is hard to see any experimental implications of the theory with respect to how matter actually behaves in relation to space, but it is still a substantive theory about what matter and space *are* and how they are related. Experimental and theoretical physics need to be supplemented, I believe, by what can only be called a priori physics, however offensive to contemporary sensibilities such an enterprise might sound. Descartes and Locke were engaged on a perfectly serious enquiry, even if it is properly described as metaphysics—or meta-physics.

---

[31] Regarding the nature of space as defined and exhausted by a mathematical coordinate system will certainly make one blind to the question, since space will then be conceived merely as a triple of numbers. Such a mathematization of space will result in simply assuming that space exists inside matter, because a suitable coordinate triple of numbers will be assignable to the place where matter is; there will then be no further question of whether space exists at that location. The mathematician's abstract notion of a "space" has taken over from the original idea of concrete space, so that my question becomes invisible. To get a feel for my question, you need to re-acquaint yourself with space as a concrete reality and forget the mathematician's special sense of "space." Also, scientists will be apt to ask how one might set about empirically verifying whether space exists inside matter (does one have a look inside?); not seeing any answer, they will dismiss the question as empty. But I am doing metaphysics, and here empirical verification is beside the point. I am indeed asking about a question of fact, but there is no direct way to settle the question empirically—we must appeal to considerations of logical coherence and explanatory power.

Postscript to Chapter 1
Particles as Fields: An Objection

I more or less presuppose in chapter 1 that matter is impenetrable, and then I try to provide an explanation of the impenetrability. But it may be objected that my *explanandum* rests on obsolete physics—the Newtonian physics of the hard bounded particle in a fixed location. What if we update this conception to construe a particle, such as an electron, as a centered field of force, with no solid kernel, just energy radiating from a central mathematical point? Then, such impenetrability as we observe results from the repulsive force produced by such a field—specifically, electrical force. But this repulsive force is not absolute: it comes in degrees, and in principle particles so construed could overlap in space, because fields can overlap. Metaphysically, two electrons *could* exist in the same place at the same time if forced strongly enough together. Field-based physics thus allows for metaphysically possible overlap, and hence denies impenetrability. Thus there is nothing to explain by means of the Deletion Theory—there is just ordinary electrical repulsion.

Let me distinguish strong and weak impenetrability: the strong kind is that presupposed in chapter 1 and advocated by Descartes and Locke (among many others); the weak kind is that acknowledged by the objector just cited—impenetrability that comes in degrees and is not absolute. Why do I believe in strong impenetrability and not just in weak? The strong kind is implied by the metaphysical principle that there cannot be two things in the same place at the same time (unless one composes the other, as with the statue and the piece of bronze). But why believe that principle? Is it just that we have never actually seen such a thing happen? But that merely observational fact is explained by weak impenetrability, given the strength of the forces we know to be at work in nature. The correct answer comes from considerations of *individuation*: we don't know how to *count* objects if strong exclusion is denied. Suppose it is allowed that two electrons could be spatio-temporally exactly coincident, by dint of the overcoming of mutually repelling forces: but why should we say there are *two* here? Add a third electron to the mix—why do we have three in one place now? Why not say instead that three electrons have become one, since they are quite indistinguishable once they precisely share their location? Admittedly, the resulting particle will have three times the mass of the particle joined by the other two—but why not say that the three electrons have combined to form a particle of triple the mass of an ordinary electron?

Don't reply that electrons are individuated by their mass because we can always re-describe the situation by saying that a *new* type of particle has been created—which is not an electron, but has three times the mass of an electron. The question is what justifies the assertion that we have three *electrons* sharing the same location, instead of a single heavy electron or a new type of particle. Indeed, there is nothing to prevent us, if we accept only weak exclusion, from allowing that *every* electron could occupy the same place at the same time, as a matter of metaphysical possibility: but then what is the ground for describing the situation that way, instead of saying that all the electrons have joined to form some sort of massive super-particle combining all their masses? We individuate physical things by their spatio-temporal coordinates, so we lose the sense of identity and difference if we deny strong exclusion.

And the same point can be made at the macroscopic level: weak exclusion allows cats and stars to merge and yet retain their identity, as a matter of principle. Suppose each particle of such an object to move into the same position as a corresponding particle of an object of the same type (a twin cat or star): we will then have a cat-like or star-like object made up of the particles of the original pair of objects, with total particle overlap. So do we now have two cats or stars that are spatio-temporally coincident? But why not say, more plausibly, that we just have a single cat or star that has doubled its mass? Individuation has broken down. In fact, if we deny the metaphysical principle behind strong exclusion, there is nothing to stop us from saying that every physical object now existing is really the amalgam of indefinitely many still-existing objects—that every electron or cat or star can be described as *many* such objects that happen to be spatio-temporally coincident. Individuation goes haywire once we deny strong exclusion.[1] The case is just the counterpart of denying

---

[1] Notice that the same individuation problems afflict other denials of strong exclusion—for example, of regions of space, events, and selves. Why can't we allow for the possibility that two or more volumes of space should be spatially coincident? Why do we insist that a given region holds only a *single* volume of space? Because otherwise all constraints on the individuation of places collapse: we might as well say that a given region contains seventeen volumes of (qualitatively identical) space as say that it contains one, or infinitely many. Similarly for events: could two events of the same type be spatio-temporally coincident—say, two wars? But what would make them two and not one or seventeen? Spatio-temporal location is individuative. Same for selves: two selves merged imperceptibly together or one new self or seventeen? If your brain and mine merge, by systematic

the principle that objects in *different* places cannot be identical: if we allow that objects in different places can be identical, we can no longer distinguish object identity from object (exact) similarity—we might as well describe many replicas as a single particular existing at different places.

It might now be thought that physics evidently regards as nomologically possible what is metaphysically impossible, since physics (allegedly) treats particles as in-principle penetrable fields. That is not a happy outcome, since metaphysical modality is by definition more lenient than nomological modality. We don't want our physics to countenance logical contradictions! But I don't think we are compelled to interpret physics in such a way as to lead to that kind of outcome. The key word here is "interpret": empirical findings are *interpreted* as suggesting that particles are really penetrable fields—the whole idea of the field is a theoretical construct. The lesson of quantum theory, however, is just that we don't *know* what is going on down there; and the concepts we bring to bear are inadequate tools with which to comprehend the phenomena—including the concept (or image) of a resistant field of force. What we don't want to end up doing is interpreting the experimental data in such a way as to countenance violations of strong exclusion, on pain of making a complete hash of individuation—not unless we really have no theoretical alternative (including the admission that we have no idea what might be going on at the microscopic level). The idea of particles as tiny resistant fields of force, where the resistance is graded and never absolute, is not a *datum* of physics—it is just one interpretation of obscurely perceived reality. But it really is a datum of metaphysics that there cannot be two (or more) distinct material things in the same place at the same time (unless one constitutes the other).

Moreover, it is not clear that this theoretical model of underlying physical reality, even on its own terms, allows for total overlap: for what about the *center* of the field—the point lurking in the middle? Resistance increases as we approach that center, conceived now as a mathematical point: one central point approaches another,

---

particle co-occupation, is the result one self or two or more? The alleged merging of twin brains can be described either as one self that results from two, or two selves that appear as one, or neither—the whole question seems moot. Lesson: don't allow that two brains *could* so merge. Surely no one will maintain that physics entails that *all* entities could so merge and yet retain their pre-merger identity, so why insist that it entails that for the special case of particles or bits of matter? We really don't want to allow for the metaphysical possibility that every particle in the universe could robustly exist in the same place as some single such particle—as it might be, the whole universe of particles overlapping with a particular particle now in one of my fingernails. I don't mean all the *material stuff* of the universe condensed into that tiny area, as with the big bang scenario—I mean every particle, as it now exists, fully itself, existing in that area, with no competition for location. There would be no *crowding*, according to such an overlapping model, since particles are held not to compete intrinsically with others for location. If particles were not electrically charged, as neutrons are not in the actual world, then there would not, according to the model in question, be any physical obstacle to total overlap, so that the laws of nature would not preclude such overlap—like those coexisting angels of the Middle Ages dancing on the head of a pin. But matter is more exclusive than angels, essentially so.

encountering stronger and stronger resistance. But doesn't that resistance approach infinity as the two centers asymptote towards each other? If so, the repulsive force is infinite when they reach the point of complete overlap—which is to say that it cannot logically be overcome. So we get strong exclusion with respect to the central point, and hence the fields cannot completely coincide—they can only approximate to total overlap. And if that is so, then the principle that matter essentially and absolutely excludes is preserved, even if we accept the field interpretation at face value.

How does the field interpretation of particle talk fit with the Deletion Theory? Not happily, it would seem, since fields can apparently occupy the same existent spatial region, as I note in chapter 1. Fields do not delete the space over which they operate. But if there is nothing to matter *except* fields, then matter does not delete either—so the Deletion Theory of matter in the actual world would appear to be empirically false. Particles and the objects they compose, being ultimately field-like, do not do anything so drastic to space as to remove it from existence. So does this mean that under the pure field interpretation of physics we must abandon the idea of matter as an operation on space?

Not necessarily. True, the operation cannot be simple deletion, because that would rule out the spatial overlap of fields, for the reasons given in chapter 1. But there is room for less dramatic operations on space, as I also note: why not say that fields, though they do not *delete* space, can nevertheless *dilute* it? Call this the Dilution Theory. According to this theory, when a field comes to pervade a region of space it *thins the space out*, so that it is less "dense" than before. Being less dense, so that there is less of it per unit volume, the usual total penetrability of empty space is compromised—so that it becomes more resistant to occupation. In other words, graded resistance, such as that generated by an electrical field, rests upon, and is explained by, the Dilution Theory. While solid matter deletes space (or would delete it if there were any), and hence makes it unavailable for further occupation, electrical fields act to dilute space, and hence make it *harder* to occupy (by like charged particles). In its most radical version, the Dilution Theory says that the existence of a resistant field *consists* of diluted space—as solid matter consists of deleted space, according to the radical Deletion theory. Thus matter is reduced to deleted space and force is reduced to diluted space (or space-time): both physical realities are conceived as reducible to modifications of space—in somewhat the way General Relativity reduces gravity to spatial curvature.

My point here is that the conceptual apparatus developed in chapter 1 is flexible enough to allow for this kind of theoretical extension. And empirical underdetermination of theory permits the formulation of such (bizarre-sounding) theories. That is, the pure field interpretation of particle physics, in which there is nothing but graded force, is not *inconsistent* with the basic conceptual framework sketched in chapter 1. We just have to replace deletion by dilution, in order to accommodate the weaker kind of exclusion contemplated by the notion of in-principle penetrable fields. Not that I think we should do that: as I suggested above, I don't think this kind of interpretation is logically required by the experimental facts and it leads to intolerable problems with individuation. But *if* you are wedded to such an interpretation, you need not abandon my general framework of conceptualization; you just modify it accordingly. Empty space still differs from matter by virtue of its total penetrability, but matter is now defined as what is impenetrable *to some degree*, instead of absolutely. Descartes and Locke regarded material bodies as absolutely impenetrable—an all-or-nothing matter—but we can still employ their essential insight if we prefer the notion of different strengths of resistance. Matter is still essentially different from space in offering *some* resistance to motion with respect to other matter; its essence is still impenetrability—just graded impenetrability.

Thus we can continue to hold to the spirit of the Lockean definition of matter, as well as accept an amended version of my space-modification theory of spatial occupation, also suitably weakened (partial dilution not complete deletion). I prefer to follow the simpler, more absolute, version of the theory developed in chapter 1, but the basic framework can be modified to fit interpretations of physics that seek to reduce matter entirely to interpenetrating fields—whether or not this is a compulsory, or even a good, interpretation. The usual motivation for such an interpretation is typically by appeal to Occam's razor: if we can get by with fields alone, and we need fields anyway, why not banish anything else from the theoretical structure? Why work with a duality of particles *and* fields? I am suspicious of this kind of motivation in general, since there is no guarantee that nature shares our hunger for simplicity; but even if you feel the need to succumb to it, you can still enjoy the benefits conferred by the framework I develop in chapter 1 (assuming you are moved by those putative benefits).[2]

---

[2] This postscript resulted largely from valuable discussions with Peter Lewis. Galen Strawson's comments were also a stimulus.

Appendix 1
The Uniformity of Matter

A natural view of the physical world, which was embraced by the Greek atomists, runs as follows. Bodies, large and small, consist of matter—particles as well as composite objects. Matter is a uniform homogeneous substance that comes in varying quantities. A given body contains a certain amount of this substance, which may differ from the amount contained in another body. Such differences determine the mass of the body in question, so that bodies vary in mass according to the amount of matter they contain. The quantity of matter in a body is not the same as its spatial volume, since matter can come in different densities: bigger objects can contain less matter than smaller ones, and hence have less mass. This can occur because, despite the uniformity of matter itself, bodies can vary in the proportions of matter and space within their boundaries: the particles of matter that compose a body can vary in the distances between them, so that one object can contain the same quantity of particles as another and yet these particles may be spatially more separated, thus producing an object of larger volume. The density of matter turns on the spatial proximity of the constituent particles: the closer they are, the denser the matter. The particles themselves are not more or less dense—they are uniformly constituted—but collections of them can vary in density according to their spatial relations. Very dense objects, like black holes, have tightly packed (uniformly constituted) particles, while diffuse objects, like gases, have widely separated constituent particles. Thus volume and mass are independent magnitudes, consistently with the uniformity of matter.

Objects do not all have the same mass per unit volume, though matter itself does, because matter can vary in density, where density is a function of spatial separation.

A smaller object can have greater mass than a larger object because it squeezes more particles into a smaller region of space, by placing them closer together. Matter is always and everywhere the same substance, on this view, but it can be packed more or less tightly. Heisenberg, endorsing the ancient thesis of what he calls the "unity of matter," writes: "All elementary particles are made of the same substance, which we may call energy or universal matter; they are just different forms in which matter can appear"(134).[1] In much the same way, space is always and everywhere the same, homogeneous and uniform, even though it can be more or less fully occupied by matter. The substance of the physical universe is thus entirely singular: when we refer to it with the word "matter" we refer to a single natural kind, although one of extreme generality. This constitutive homogeneity fits with the dispositional homogeneity of uniform impenetrability: if it is always the same stuff, its equal impenetrability is assured. In terms of the Deletion Theory, uniform matter will perform an identical deletion operation on the space it occupies: the resulting holes are uniform. When God created the universe, he was very parsimonious in his choice of basic building material—creating just one type of stuff to make everything. Or again, the primordial material of the big bang was all of a type. The variations we observe in nature never result from heterogeneity in the basic substance but only in its pattern of aggregation—form not content.[2]

But this natural and pleasing picture encounters a problem from modern physics. According to accepted physics, particles themselves can vary in mass, sometimes quite dramatically, and yet there is no empty space within them. Electrons, protons, and neutrons all have different mass. Measures of inertia reveal that particles can vary in their degree of inertia, and inertia is the standing definition of mass; but mass is the amount of matter contained in a body, so that particles can differ in the amount of matter they contain. Suppose that a particle a has a mass m and that a particle b has a mass $2m$; then, obviously, the amount of matter composing b is twice the amount composing a. But this difference clearly cannot be explained by supposing that the particles in a are more widely separated than those in b, since these are, by hypothesis, elementary (indivisible) particles; it is not a matter of a tighter packing of parts. It might be thought that b is simply larger than a, i.e., occupies a greater spatial volume, in which case matter still has a constant mass per unit volume, and hence is uniform.

But there is no reason in physics to suppose that there is this correlation between mass and volume at the level of particles, any more than at the level of macroscopic

---

[1] Werner Heisenberg, *Physics and Philosophy* (Harper Perennial: New York, 2007).

[2] A similar assumption might underlie the belief in mental substance: just one type of mental substance for all the many types of mind—different species, different sexes, old and young. According to Cartesian dualism, the mind stuff would differ from the matter stuff in its essential nature, but on either side of the divide uniformity would prevail. Of course, it would also be possible to maintain that a single stuff encompasses both mind and matter—spiritual, corporeal, or neutral.

objects: nothing in the laws of physics prevents a and b from having the same size (spatial extent), despite their difference of mass. The physicists are strangely silent, or guarded, on the question of the radius of elementary particles, focused as they are on their mass, motion, charge and spin, perhaps because radius plays little to no role in fixing the behavior of particles: but it is not precluded that mass and radius are independent variables. Let us suppose then that a has the same spatial extension as b, despite differing in mass. How is that possible? It can't be that the parts of b are more compressed in space than those of a, since these particles have no parts and contain no space; so the apparent difference of density does not work in the way it does for large composite objects. Then how does it work?

We seem to have just two options. One is to hold out hope for a conception of density that does not require interstitial space—the idea, in effect, of reducing the volume of a continuous substance. But it is not clear what this could be: once we reach continuous matter it has done all the reducing and compacting it can do. We might try the idea that it can somehow fold in on itself, but this commits us to the consequence that distinct units of matter can occupy the same place at the same time—not a happy result. It also commits us to the idea that elementary particles have a more complex structure than we thought, though we have no empirical evidence for this. Also, such compression or folding would appear to require a force to bring it about—as gravity causes the compression of matter inside a black hole or kinetic pressure can compact a car. Yet there is no evidence that more massive but equally sized particles are subject to such compressing forces: they have greater mass than others of their size on their own account, without benefit of some supposed compacting force. So it is hard to maintain that a and b are composed of the very same stuff—matter *tout court*—which is merely denser in one than the other. Given that particles are composed of matter, differences of mass at the particle level therefore pose a problem for the unity of matter thesis, on the plausible assumption that size and mass are not correlated.

And so we reach a possible revision in our general conception of matter: perhaps fundamental matter comes in different *types*. I don't mean the trivial proposition that there are different types of elementary particles—protons, electrons, neutrons, etc.; I mean that what *composes* these particles might be different types of *stuff*. Some types of stuff will inherently have more mass than others per unit volume, and this is why a and b differ in their mass—because, in short, one is made of heavier material than the other. It is not the same substance differently compacted that accounts for the difference of mass, but totally different types of substance. We are driven to the admission of the *disunity* of matter. Instead of operating under the assumptions of ancient atomism, according to which all atoms are composed of the same substance (though they can have different shapes and sizes), we are forced to revert to a characteristically pre-Socratic thought, namely that the world consists of a number of irreducibly distinct types of composing substance. Not earth, fire, water and air, to be sure, but something logically akin to that—substances that are fundamentally and irreducibly different in their inner nature, not variations on a single theme. What we call by a single name—"matter"—is really a disjunction of different stuffs; there is no single natural kind of stuff that we are referring to. There are varieties of matter; or better, the word "matter" ambiguously stands for several distinct natural kinds.

And let me repeat, I am not speaking here of the different types of *objects* distinguished by particle physics—of course *they* come in different natural kinds (electrons, protons, etc.); I am speaking of the primordial stuff that *composes* them ("matter" as a mass term denoting a stuff). It is entirely consistent to suppose that particles come in different natural kinds even though the stuff that composes them is of one kind; what I am contemplating is the more radical move of questioning the unity of the underlying substance—whether we call it "energy" or "matter." On this supposition, some kinds of so-called material stuff could simply have greater mass per unit volume than others, which is why the elementary particles differ in mass. Protons and electrons might be composed of different *types* of "matter," as well as differing in charge, spin, etc.[3]

What is remarkable about this suggestion is (a) that it is prompted by a basic fact of physics, viz. particles having different masses, not a metaphysical conjecture; and (b) that the indicated consequence, viz. different types of fundamental matter, is so radical and unconsidered. For it is not as if physics has embraced, or even seriously considered, adopting such a view: the discipline remains studiously silent on the real nature of matter itself (as opposed to the behavior of the entities it composes).[4] But the metaphysics of physics cannot help confronting the question: is everything "physical" made of the same stuff or not? Does the physical world bottom out in a homogeneous substance or does ontological variety go all the way down? There would be no obstacle to the former view if elementary particles all had the same mass (or their size and mass were suitably correlated), but that seems not to be true; so we are forced to consider whether matter is really the uniform substance that we tend to suppose. And if it is not, how many types of matter are there? How do they differ? How might they affect the laws governing particles? Do the differences show up at the macroscopic level? If not, why not? How might we set about empirically answering such questions? This is all very difficult.

I have to admit that abandoning the uniformity of matter goes against the grain for me: it seems far more appealing to me to suppose that all massive bodies are composed of fundamentally the same substance. The plasma of the big bang is generally thought to have been uniform, before particles evolved. Unless we are to suppose that particles really consist of nothing—that the question of what composes them is somehow illicit—we are faced here with a difficult issue, and no obvious way

---

[3] Or different types of energy—if we follow Heisenberg and other physicists in making energy basic.

[4] See chapter 2 on why physics doesn't reveal the ultimate nature of matter.

to resolve it. Granted, all matter will share certain general properties, such as occupying space or giving rise to inertia; but the prospect of an irreducible heterogeneity in the bedrock is sufficiently contrary to our general worldview that accepting it amounts to something on the order of a paradigm shift, or apple-cart upset—and not one that leads to greater theoretical simplicity or predictive power. The pre-Socratic theories succumbed to the atomist theory partly because of its unifying power—the same basic stuff occurring (or recurring) in different forms—but now we are contemplating reverting to the notion of a primitive variety in nature: stuff pluralism, not stuff monism. The problem here is that we really have no idea of what matter is anyway, aside from its dispositional or structural properties, and now we find that we don't even know whether it is uniform or mixed. We have uncovered some *prima facie* reasons to doubt the unity thesis, but how we might settle the question is extremely difficult to say. We are thus ignorant of one of the most basic properties of matter—whether it is uniform or not. And this might well be one of those "mysteries of nature" that we will never crack.[5] Even if we could rationally come to the view that matter is not uniform, we might be precluded from further grasping the inner nature of the variation: we might grasp indirectly and abstractly that it is so and yet not be able to get a grip on what the variety consists in.

The panpsychist, always eager to seize her opportunity, will doubtless step in to explain that *she* knows what the varieties of matter consist in—they are simply the varieties of consciousness: electrons being constituted of sensations of red, say, and protons constituted of sensations of blue.[6] The irreducible varieties of consciousness will then be paraded as constituting the irreducible varieties of matter. More reticent souls will confess to a deep ignorance about the intrinsic nature of matter, conceding that we don't even know enough to say decisively whether matter is unitary or multiple—or what this difference would even come to. What is alarming is that a seemingly elementary fact of physics—that particles differ in mass—can open up such an abyss of uncertainty and metaphysical anxiety. The ignorance is quite close to the surface.[7]

---

[5] The quoted phrase is Hume's. See the Introduction, note 6, for the source.

[6] I say this not to parody the panpsychist, for whom I have a deep (if grudging) respect, but only to make vivid how she might account for variety at the most basic level: she at least has a concrete idea of what the different types of matter might consist in. No doubt a practicing panpsychist will speak more cautiously (or darkly) of different basic types of "proto-consciousness" as constituting the varieties of matter.

[7] Much more will be said about our ignorance of the physical world as the book proceeds. The question of the uniformity of matter illustrates the general point well, and shows how physics can raise difficult epistemological and metaphysical questions quite quickly.

Appendix 2
Divisibility and Size

The physical world divides into the divisible and the indivisible. It is an interesting fact about our world that the indivisible things are very small. There is clearly no analytic or conceptual connection between divisibility and size: there is no contradiction in the idea of indivisible but large objects. That is, there could, as a matter of conceptual possibility, be elementary particles the size of the sun. Presumably, such a particle would have a mass many times greater than the sun, granted some assumptions about the uniformity of matter and the lack of space inside particles. But nothing in logic or physical theory precludes the existence of indivisible particles the size of the sun.[1] It appears to be merely contingent that the indivisible things in our world are also the small things.

It is not contingent that the divisible things are bigger than the indivisible things that compose them (if they weren't they would be bigger than themselves!), but smallness of scale is not a necessary concomitant of indivisibility. It has turned out that the indivisible things are unimaginably small compared to what the ancient atomists suspected, but nothing really rules out having discovered that the indivisibles are detectable through a magnifying glass. This is not to say that quarks, for instance, are only contingently tiny—that *they* could have been as big as the sun; arguably,

---

[1] When I say "in physical theory" I don't mean that such big particles are nomologically possible in our world (their mass would be absolutely enormous). I just mean that the theory of particles as having mass, charge, spin, location, and so on, does not preclude very large particles. Nothing in physics requires that indivisibility be necessarily correlated with the smallness of scale we observe in our particles.

they wouldn't then be *quarks*. It is only to say that it is contingent that the world contains quarks and not, say, mega-quarks which are as big as the sun—capable of composing even bigger objects, perhaps, but not themselves composed of smaller objects. Of course, in such a world we would likely not have the same kinds of macroscopic objects that we do now (e.g., galaxy-size ants composed of mega-quarks), but there might be other kinds of enormous objects that these massive particles composed. In that kind of universe the Deletion Theory tells us that sun-sized holes are cut in space by the continuous matter of these colossal particles, instead of the tiny holes cut by our actual minute particles.

This reflection may seem surprising—after all, the smallness of the indivisibles is, to put it mildly, a remarkably pervasive trait of our actual universe. It hardly seems like just a fluke that there aren't any very large particles around. But the surprise is somewhat mitigated by the fact that size—or our sense of it—is an anthropocentric phenomenon. From whose point of view is an electron tiny? From whose point of view is the universe vast? Such descriptions reflect our own dimensions and the way we carve up the universe correspondingly. There is really no absolute sense in which a galaxy is extremely big and an electron is extremely small: such judgments of size are relative. If you were a being half the size of an electron, it would seem pretty large to you; and if you had a head as big as a galaxy, a galaxy wouldn't seem so vast. To say that a particle is tiny *simpliciter* is meaningless; it is the size it is, that's all. What we really should be saying is that it is contingent that indivisible things are not as large as the divisible things we now observe (and maybe some of the larger mass elementary particles are actually bigger than composites of lower mass particles—though the relevant physics doesn't seem to bother with questions of size). After all, the sun is tiny compared to the visible universe—as the electron is tiny compared to the sun. Compared to the infinite everything is vanishingly small.

You may balk at what I have just said: how, you may wonder, could a particle be as big as the sun and still be indivisible? It will be distributed over the same region of space as the sun (by definition) and hence admit the same divisions that such a large region of space permits. That is, the matter composing such a "particle" can be divided into distinct spatial segments, geometrically speaking. Indeed, the big particle will be composed of a huge number (an infinite number) of "point particles," corresponding to the spatial points that fall within the particle's boundaries—any occupied point of space, in effect. In other words, the material stuff that composes the big particle will be spread out over a large region of space and can accordingly be sliced into many parts. In this sense, we must agree, it *is* divisible. Notice that the same reasoning applies to smaller but finite particles—those that make up actual matter: the big particle just dramatizes the point. Does this mean that point particles are the only genuine indivisibles—the only things that admit of no division? That is, are infinitesimally small units of matter the only genuine atoms?

The answer is that it depends what you mean by "indivisible." The sense just in play might be called *geometric* divisibility; it corresponds to the geometric divisibility of space. Any finite particle is divisible in this sense, as a matter of mathematical necessity, just as (continuous) space is. Clearly, the only indivisibles will be infinitesimal points on this understanding (will lack size, in fact). But this is not the only notion that can be defined, or the most useful. I think there are two other distinguishable notions of divisibility, and it aids clarity to mark them out. One is the idea of spatial dispersal—in brief, *cutting*. When an entity can be cut into separable components, and these components can be sent in different directions, we have divisibility in a very robust sense. In this sense, it was discovered that the atom is divisible, since electrons could be split off from the nucleus and these take different paths through space. Actual physical experiments are needed to establish this kind of divisibility, not mere geometrical reasoning. And obviously the possibility of cutting does not follow from mere divisibility of spatial extent (the geometrical notion): a particle might have geometric parts that cannot, as a matter of physical necessity, be separated from each other—indeed, this seems precisely true of our elementary particles.[2] Clearly, too, it is no argument for point particles as the only indivisibles to insist that anything bigger than a point is divisible and hence not a particle: a particle might be divisible in the geometric sense (and hence not a mathematical point) but not in the cutting sense. It is then merely a verbal quibble to argue that such a particle is really divisible: the proper response is—in what *sense* divisible?

However, in one respect the cutting definition is too strong: we can identify a nontrivial notion of divisibility without requiring the physical possibility of spatial dispersal. We can call this third notion *functional* divisibility: that is, an entity might be divisible in the sense that it can be divided into differently functioning parts and yet these parts cannot be physically detached from each other. Conceptually, the point is easy to grasp: an animal body, say, is divisible into functionally separate parts—arms, head, tail, etc.—but it does not *follow* from this that the parts can be physically detached (though of course they can as a matter of fact be detached). Similarly, there might be a physical entity that has functionally distinct parts—parts that do different things and obey different principles—and yet it is not possible to break it apart and send the parts away. If String Theory is true, electrons may be made up of multiple strings that vibrate differently from each other, and yet there is no physical way to break the electron into these strings. Breakability is a matter of physical forces and whether other forces can overcome them, and this is quite different from the idea of having an articulation of distinct functioning parts. Crudely, the distinct functional parts may be (indissolubly) *stuck together*. This third notion is also distinct from the geometric notion, because that notion requires no division into functionally distinct parts—the parts may be utterly homogeneous, like sub-regions of a region of space.

---

[2] Actually, it is arguable that the impossibility of cutting continuous matter is logical, not merely causal—because cutting requires some interstitial space through which the knife moves. The absolute impenetrability of continuous matter prevents anything from passing through it, which is what a cutting agent would have to do. There is no logically possible process of cutting by means of which the (geometric) parts of a continuous substance could be separated.

So it would be a fallacy to argue that a particle must be divisible in this (robust) sense simply because it is divisible in the (trivial) geometric sense. Accordingly, an elementary particle the size of the sun would be divisible in the geometric sense but not in the other two senses; so there would be two good senses in which it really is a particle, an *atom* in the original acceptation.

Conceptual confusion can result from not distinguishing these three senses of "divisible." Thus someone might maintain that strings cannot be basic because, being finite in length, they divide geometrically into parts; that may be true, but it doesn't follow that they are not basic in the other two senses—which are far more physically significant. And I suspect that the usual talk of point particles comes from taking over the geometric notion of a point as that which is basic in the construction of geometrical figures—the geometrical notion of an atom, in effect. But this is a far cry from the kind of atom needed in physics, which has more to do with lack of functioning parts than being a mathematical point. We can agree that any finite chunk of matter is made up of points, but it is confusion to infer that these points are the only genuine atoms. An indivisible particle can indeed have many (infinitely many) point particles composing it; this is no contradiction, but a reflection of the different notions of divisibility that might be in play. It is also acceptable to use point particles as idealizations of actual particles for mathematical purposes, but of course it doesn't follow that actual particles *are* such points. Anyone troubled by the idea of point particles having nonzero mass will want to keep these distinctions in mind.

Once these distinctions are kept clear, the idea of a universe with particles of arbitrarily large size seems not to pose any conceptual problem. There could have been atoms as big as a galaxy. Matter itself seems to impose no restrictions on how large its indivisible units may be, at least as we now conceive it. In fact, the existence of divisible units of matter is not a consequence of the very nature of matter either: there could have been a world containing *only* indivisible units of matter, small or large. Our world is thoroughly populated with divisible physical entities, but the essential nature of matter is not responsible for this fact. Matter could just as happily exist without forming such units—every entity could be indivisible and matter would be the same stuff it is now (though it might be subject to different laws). Nor is there any reason stemming from the nature of matter why it should not form only indivisible units of arbitrarily large size. It just so happens, in virtue of local conditions, that objects tend to be composite and the atoms composing them are small.

Compare the state of water: the thermal conditions that obtain on earth render most water liquid (at least where the majority of people live); we therefore naturally assume that this is somehow the natural condition of water, so that its frozen and gaseous forms are departures from the norm. However, nothing in the very nature of water supports any such assumption; and in different thermal conditions water may exist only in a frozen state or gaseous state—the liquid form would seem anomalous for those living in such conditions. I am suggesting that the specific configuration in which matter typically appears to us—in respect of size and divisibility—is likewise parochial. In its intrinsic nature, matter is neutral as to the scale of its most basic units and their participation in the formation of composite objects. If we lived in a world in which matter always came in huge indivisible chunks and never combined to form bigger objects, then we would form a very different idea of its essence—but it would be the same old matter with which we are now familiar. Or to put it more cautiously, nothing we *know* about matter now is inconsistent with supposing that it could exist in this very different form.

And I don't mean that some stuff *similar* to matter (an epistemic counterpart of it perhaps) could exist in such a form; I mean *our* matter could be like this. Matter, after all, can make up many kinds of particle in our actual world—the same stuff recurring in different forms—and so there is no bar to its making up even more types of particle in a possible world, such as an indivisible unit the size of the sun. It may well be that the particular conditions obtaining at the time of the big bang determined the specific structure that matter assumed—small particles prone to form composite wholes—but that these conditions are just one of a spectrum of physically possible conditions, others of which might have led to very different material formations. It is often acknowledged that the specific values of basic physical magnitudes are highly idiosyncratic and arbitrary; perhaps the very structure of the material world that we see is likewise idiosyncratic—tiny combinatorial atoms being just one rather eccentric way that matter could turn out. All matter must have extension and impenetrability, but that it should be divisible into microscopic constituents looks like a contingency. Spatial occupancy is what matter is all about, and this dictates neither microscopic particles nor divisible composites.

But let me emphasize that this conclusion rests upon reflection on what we *know* of matter—our current conception of it. Given, however, that our knowledge is partial, and possibly distorted, things might be quite otherwise objectively. It *could* be that matter has a nature such that there *is* a definite bound on how large an indivisible unit of it can be—so that sun-sized particles are in fact a metaphysical impossibility. That would certainly make sense of the remarkable agreement in scale of the elementary particles we observe. The contingency as to size that I have suggested *might* be a symptom of our ignorance of matter. Giant particles certainly seem like real possibilities, but it is hard to be sure.[3]

---

[3] As in other cases, ignorance of a thing's nature can give rise to illusions of contingency. The big particle *might* be an epistemic possibility without being a genuine metaphysical possibility. It is certainly not contradictory, but something about the inner nature of matter might put an upper limit on particle size. Still, my sense is that big particles are a real possibility, because in general size is so contingent: things of the same form can come in different sizes—for what has form got to do with magnitude of extension? We see things of the same nature with different sizes all the time.

2 What Is a Physical Object?

The traditional conception of a physical object analyzes it into three levels or aspects: primary qualities, secondary qualities, and substance. The primary qualities include extension, solidity, motion and number; these are deemed mind-independent and intrinsic to objects. They constitute the nature of the object as it is in itself, and they are bound up with the laws that govern the behavior of the object. The secondary qualities include color, sound, smell, feel, and taste; these are often deemed mind-dependent, and strictly speaking extrinsic to the object. They constitute the appearance of objects to the human senses, and they are irrelevant to the physics of objects—particularly, their dynamical properties. In addition to these two levels, there is the matter that composes the object—its substance. Objects *have* primary and secondary qualities, but they are *made* of matter. Extension is not itself matter; rather, extended bodies are composed of matter. It is what composes an object that properly speaking makes it material, as distinct from the various primary and secondary qualities it has. Strictly, then, it is not the primary qualities that make an object material; it is the type of stuff of which it is made. To be a physical object is to be made of material stuff—whatever that may be.

This may be questioned. Why do we need to add the third level to the first two? Can't we get by with just the primary (and secondary) qualities? Can't an object just *be* the bearer of such qualities as extension, motion, solidity, and number? It is important to see why that is not possible. The essential notion of matter is what takes up or occupies space—of that which is extended *in* space; and this is not the same as the space that is occupied. Much can be said about what this idea of taking up space involves (I discuss it in chapter 1), but for our purposes the key point is that mere extension is not the same as the thing which is extended—the material thing. Space is also extended, but it isn't matter (*pace* Descartes); so matter can't simply *be* extension. The relevant notion of extension must be the extension *of* something (other than space)—and this something is precisely a piece of matter. We cannot *identify* matter with extension, on pain of losing the distinction between matter and space. Neither can we invoke solidity to say what it is that composes physical objects. It is true that matter is solid, arguably by definition, but objects are solid (i.e., impenetrable) *in virtue* of something—it is *because* physical objects are made of matter that they are solid.

So merely noting that an object is solid does not tell us everything about it physically; something is omitted by such a description, namely what composes the object in such a way as to *make* it solid. Solidity is a relational property, defined in terms of its resistance to other objects, and it must have an intrinsic ground—that is what matter is. We need to say what *constitutes* the solid object. In short, it is not enough to be told that an object is extended and solid: that leaves out the stuff that composes the object, and matter fills the gap. An object is physical by virtue of the type of substance that composes it, viz. material substance. That substance is indeed extended and solid, but neither property constitutes the inner essence of matter—what it intrinsically *is*. Extension is too broad, and solidity too relational-dispositional. So the tradition assumes, and reasonably so.[1]

It is worth noting that the geometrical properties of matter all have counterparts in the properties of space. Extension should be taken to include shape—the specific configuration that a certain quantity of extension (i.e., volume) takes. Objects obviously come in different shapes and sizes, and their movements also trace out shapes—straight line or circular, say. Clearly shape is essential to our notion of a physical object. It might be tempting then to think that shape is the be-all and end-all—that material things are just things with shape. If so, the primary quality of shape would be enough to qualify something as material. But space too has shape; indeed, it has every shape that matter has. Within space every geometrical form possessed by occupying objects is instantiated, since it is made up of points that trace out every such form. This is precisely why shaped material objects can fit into space—because space already has the geometrical form of every object. Thus we can speak of a triangle in space, each of whose vertices is a certain point, materially occupied or not. It follows that shape as such cannot be *definitive* of matter; it is not sufficient to confer materiality. In general, the primary qualities are qualities *of* matter; they are not what *make* something matter. What makes something material

---

[1] I am thinking mainly of Locke's conception of a physical object (a "body"), but the same view was taken by many others: we can't eliminate the concept of material substance in favor of other properties like extension and impenetrability; that concept is crucial and indispensable in accounting for body.

is what it is composed of, not the geometrical form this composing material assumes.² Merely spatial properties cannot define matter *in contrast* to space.

So: what *is* matter? What is the stuff of which physical objects are made? Here we might resort to physics. Physical objects are made of atoms, it might be said, as physics has discovered, and atoms are made of protons, neutrons, and electrons. That is true enough, as far as it goes, but it doesn't answer our question; it just pushes it back. For what are the most elementary particles made of? Matter, we will naturally reply: but then we have our original question again. What is a particle made of when it is made of matter? The *word* "matter" offers little help here: it simply means "what constitutes something." Everything has matter of some sort—even the Cartesian mind—since it has a constituting substance. The word "matter" is not much more illuminating than "substance," which likewise is promiscuous as to ontological category. The word "matter," as we use it to single out a certain *kind* of substance, functions like an indexically anchored natural kind term: matter is what *these things* are made of, whatever that may be, and not *those things*. The *nature* of the referent is not thereby disclosed; we simply have a non-descriptive label.³ Does empirical physics tell us the nature of this referent? I don't think so; it merely tells us what matter *does*, not what it *is*.

Physics is structural, and it can get by without entering into the question of what matter ultimately is. As many have argued, physics leaves a blank where the intrinsic nature of matter lies, though it successfully describes the mathematical laws that govern the *behavior* of matter.⁴ If the ultimate constituents are vibrating strings, then physics will tell us what the strings do, but it is silent on the question on what they are ultimately made of. Physics says what parts things have, right down (it is hoped) to the things that have no further parts, but it does not say what it is that ultimately composes these things—that is simply left open. The question "What is matter?" is therefore left unanswered by physics. In practice, a contrast with space is assumed, but no positive characterization is offered; matter is just "whatever it is that does such and such things."⁵ If we ask after the nature of *what* it is that does these things, we will not receive an answer (either defiantly or apologetically, depending on the physicist's philosophical viewpoint). There is an epistemological gap here—a descriptive lacuna. How might we fill in the blank?

---

² It seems conceivable that some sort of non-material ectoplasm might have geometric form in some other possible world—so merely having shape can't *entail* that a substance is of the same type as *our* shaped objects. Also volumes of light can have shape, yet are not material substance in the traditional sense; and fields have configuration too—not to mention ghosts. If space is curved, it also has shape—but is not material. It is not the shape of a thing but its stuff that makes it distinctively material.

³ Galen Strawson makes this point in his "Real Materialism," in *Real Materialism and Other Essays* (Oxford University Press: Oxford, 2008), 47–48. We are close on many points, and I have been influenced by his formulations.

⁴ I am thinking of Hertz, Poincare, Eddington, Russell, and others; I cite their writings frequently in the course of this book.

⁵ A very revealing (because unsatisfactory) definition of matter is provided by Linus Pauling at the beginning of his classic textbook, *General Chemistry* (Dover Publications: New York, 1988). The first chapter is entitled "The Nature and Properties of Matter" and Pauling begins it with these

The problem has not gone unnoticed: Eddington and Russell, among others, were much preoccupied with it. Physics is incomplete, they held, precisely on the question of what matter ultimately and intrinsically is. Physics consists of mathematical models, detailing relations between physical magnitudes, but it is mute about the intrinsic nature of the entities whose interrelations it maps. It is purely "structural." Into this descriptive void they therefore inserted a bold theory—that the intrinsic nature of matter is *mental*. This is the doctrine of panpsychism, offered as an account of what physics leaves unsaid: the essence of matter is consciousness, and the physical world bottoms out in mind-stuff.[6] On this theory, or one version of it, the primary qualities of objects are sandwiched between two sets of mental properties—the secondary qualities, constituted by relations to sense-experience, and the intrinsic, unobservable mental properties that give matter its inner nature. Only the primary qualities have an inherently nonmental nature; mainly, physical objects are mental in nature, because of both their type of constituting stuff and their manifest secondary qualities. Panpsychism is often proposed as a solution to the mind-body problem, but in the present incarnation it is intended as an answer to what we might simply call the "body problem"—the problem of what matter is. The indexical term "matter" turns out to designate a natural kind whose underlying essence is consciousness, according to panpsychism. Bodies are selves, in effect, since consciousness always requires an "I" as subject.[7] And we know what selves are, don't we?

Now I have no wish to defend panpsychism as a solution to the body problem, but I think it is illuminating as a metaphysical theory that attempts to fill the gap I have

---

words: "The universe is composed of matter and radiant energy. Matter (from the Latin word *materia*, meaning wood or other material) may be defined as any kind of mass-energy (see Section 1–2) that moves with velocities less than the velocity of light, and radiant energy as any kind of mass-energy that moves with the velocity of light" (1). Notice that Pauling operates with a contrast between matter and radiant energy, so that the latter is not a species of the former. But what is most striking is that he appeals to a property of matter only fairly recently discovered (and enshrined in Special Relativity), namely that it always travels at less than the speed of light. Perhaps this gets things right extensionally, but surely it will scarcely do as an account of the *essence* of matter. It is far too extrinsic and contingent. Is it *definitionally* impossible for matter to move at light speed? Are there not worlds in which we have light moving more slowly than in the actual world (or faster)? What about a world with matter but no light? Would we have no idea what matter is if we had not discovered the peculiar properties of light's propagation? Pauling's definition is not even a competitor with the definitions offered by Descartes and Locke; it simply doesn't attempt to get at what is logically necessary and sufficient for being matter. Still, unlike the writers of other textbooks, he does at least make an effort to provide a definition, instead of airily ignoring the question.

[6] Strawson defends this kind of panpsychism in "Real Materialism" (cited in note 3) and elsewhere, having first accepted the purely negative structuralist position that physics does not reveal the essence of matter. I first criticized panpsychism in *The Character of Mind* (Oxford University Press: Oxford, 1996; originally published 1983), 33–36. I criticize it further in my reply to Strawson in *Consciousness and its Place in Nature*, ed. A. Freeman (Imprint Academic: Thorveton, 2006). A number of others have recently adopted the same view, and it has a long history. See also Thomas Nagel, "Panpsychism," in his *Mortal Questions* (Cambridge University Press: Cambridge, 1979).

[7] Strawson accepts this consequence, bravely and clear-sightedly, in "Realistic Monism: Why Physicalism Entails Panpsychism," 71–73, in the volume cited in note 3. To me, however, it amounts to a *reductio*.

identified, following others. It really does fill that gap (truly or falsely), and the interesting question is why. It is because mental concepts are not functional or operationalist or extrinsic or merely structural: we do know what consciousness is, and hence we know what is being *said* when matter is declared to have a mental nature. We know this because we are acquainted with consciousness in the first person. The theory contrasts with the type of theory discussed in the previous section, namely that matter can be defined as extension or solidity or shape. We want to ask what *has* these qualities, since they are clearly not the end of the ontological line; but with panpsychism we are told the answer to this question in no uncertain terms—conscious states are the intrinsic essence of matter. They are the meat of the matter—the ultimate stuff of the world. Indeed, since experiences require a subject of experience, we can say that it is conscious *subjects* that have primary (and secondary) qualities. There is no sense that this answer only postpones the question. If consciousness constitutes *our* nature, and knowably so, then it can also constitute the nature of the universe in general—it is the *kind of thing* that can make something what it is.

This theory implies that a complete physics would not conform to the "absolute conception," since there is something it is like to be conscious, and such subjective facts are not accessible from an objective point of view: but this is a consequence that might be swallowed if the pill were sufficiently ameliorative. At least if we follow the panpsychist we know what matter *is*! The basic stuff is mind-stuff, and mind-stuff is completely evident to us. If matter seemed like a kind of cosmic mystery meat, then panpsychism removes the mystery by telling us exactly what kind of meat matter is made of—mental meat. Here we see the deep epistemological appeal of idealism in all its forms: it removes the ontological mystery from the world, by projecting our own nature as conscious beings outward. To be is to be experiential—subjectivity rules. In the case of physics, idealism has taken two basic forms: panpsychism and extreme empiricism—physics is either about alien subjectivities or about our own subjectivity (Eddington and Mach, respectively).[8] In either form it is concerned with familiar realities—the operations and content of minds. The gap has been filled.

It is worth observing that panpsychism need not necessarily regard matter as inherently sensory in nature; it might, following Schopenhauer, take the *will* as

---

[8] To spell this out a bit: Eddington and the other panpsychists treat matter as a locus of subjectivity in its own right, possibly quite different from our own human mode of subjectivity—as alien perhaps as a bat's subjectivity. But Mach and his fellow empiricists regard matter as explicable in terms of *our* subjectivity, so that the subject matter of physics revolves around us humans as beings with minds. On either view, physics has an ultimately subjective subject matter: it is about *somebody's* consciousness—particles or people. Therefore, it does not postulate anything different in kind or category from what we are familiar with from our first-person perspective. By contrast, views that reject both these varieties of idealism or subjectivism hold that matter is like nothing we already know about—specifically, not like our conscious states. We are thus not acquainted with its distinctive mode of being. (I think both panpsychism and Machian empiricism stem, for at least some of their proponents, from the same desire, namely to avoid the radical unknown—a perfectly understandable wish, if not one to which we should succumb.)

ontologically basic. Particles would be less like perceivers than agents, on this view: they don't have sensations, but they do engage in acts of will. This might fit the active nature of matter better, with its forces and movements. The intrinsic nature of matter is therefore volition, which again is evident to us from our own case. It is not that sentience is everywhere; decision is (or at least "proto-decision"). When bodies move it is because they will to. In either case, we have an answer to the question of the intrinsic nature of matter. We have an answer because, to repeat, we know what mind is: our conception of it is not merely indexical or functional or extrinsic. If you like, mind is the "categorical ground" of the other properties of matter, transparent and familiar. Whether the usual primary qualities are deemed supervenient on this ground or logically independent of it, matter bottoms out in something of the right metaphysical category. Just as I know what it is like to be you, because we share our mental nature, so I know what it is to be a material body—because it shares my nature too. And even if the mental nature of matter is alien to me—like that of a bat—still it is the *kind* of thing to which I stand in a privileged epistemic relation. Matter is rescued from the noumenal or merely structural by being declared phenomenal.

Panpsychism is illuminating because it is a good example of what it would take to fill the descriptive gap; it is a theory of the right conceptual *type*. But it is unlikely to attract many disciples (me included); so we must ask whether anything less extravagant could do the job. Let me then go quickly through some possibilities.

Physicists might suggest that the essence of matter is mass, and mass is something additional to such primary qualities as extension and solidity. To be a material object, then, is just to have some mass or other. I see three main problems. First, this does not tell us what constitutes matter, only how much of whatever that is an object contains. Second, mass is usually understood in terms of inertia—the amount of force it would take to accelerate a body from a resting position or a state of uniform motion—and this is an operationalist definition. In effect, mass is interpreted as a dispositional concept, captured by suitable counterfactual conditionals: but we seek something intrinsic and grounding—in virtue of *what* does an object have a certain dispositionally defined mass? The quantity of matter it contains, surely—but then what *is* that? To be sure, pieces of matter have mass (typically: see shortly), but having mass is not what matter *is*—mass is a merely consequential property of matter. Third, we are told that some particles, notably neutrinos, have zero mass, while presumably being material (they occupy space and travel through it). The concept of zero-mass matter (defined in terms of inertia) does not seem contradictory, any more than infinite-mass matter does (in a certain sense space has infinite mass, because nothing can move it). To occupy space is not the same thing as needing some degree of force to be accelerated.

A related idea equates matter with energy. This can be either a straightforward identity or a claim of derivation. Suppose we say that matter just *is* energy: does that answer our question? Only if we know what energy is.⁹ We had better not define it operationally, as is usually done, but then we are left without any account of what *constitutes* it. Energy is just as enigmatic as matter, being known only structurally.

A different tack appeals to the concept of consciousness, but in a negative way: matter is what is *not* conscious—what there is *nothing* it's like to be. The essence of matter is its *lack* of awareness, its utter mindlessness. We take the concept of mind as understood and then use it to define matter as what is *not* mind. Elsewhere I have defined consciousness as a type of knowledge;¹⁰ well, matter is what does *not* know anything (it has no *acquaintance* with anything). That sounds all well and good, and definitely on the side of truth. It is the opposite of panpsychism, of course, and it captures the contrast between matter and mind that the tradition endorses. But it is no answer to our question, simply because it is too negative: it doesn't tell us what matter *is*, just what it is *not*. Also, it arguably makes brains *not* material, because of their close connection to consciousness. The suggestion contains no positive conception of the nature of matter: in virtue of *what* does matter lack the characteristics of consciousness? We can no more define matter as what is not mind than we can define it as what is not space—though both statements are true.

At this point desperation might set in: we need to reach higher, to get more radical. What about the Spinozistic idea that matter is *God*? Maybe pantheism is the answer to our troubles. Spinoza identified God with Nature, making God immanent in things; so perhaps the stuff of physical objects is God-stuff, presumably in some sort of disguise. Instead of identifying matter with humanlike mind-stuff, we can identify it with divine God-stuff: physical objects are composed of what composes God. Put more sedately, the substance of the world is the substance of the world's creator. After all, if God is everywhere—as pantheists insist—then he is in the very marrow of the material world, and hence he provides an answer to the

---

⁹ Here is a nice passage from Clerk Maxwell summing up the epistemology of physics as he sees it: "All that we know about matter relates to the series of phenomena in which energy is transferred from one portion of matter to another, till in some part of the series our bodies are affected, and we become conscious of a sensation. By the mental process which is founded on such sensations we come to learn the conditions of these sensations, and to trace them to objects that are not part of ourselves, but in every case the fact that we learn is the mutual action between bodies. This mutual action we have endeavoured to describe in this treatise. Under various aspects it is called Force, Action and Reaction, and Stress, and the evidence of it is the change of the motion of the bodies between which it acts. The process by which stress produces change of motion is called Work, and, as we have already shown, work may be considered as the transference of Energy from one body or system to another. Hence, as we have said, we are acquainted with matter only as that which may have energy communicated to it from other matter, and which may, in its turn, communicate energy to other matter. Energy, on the other hand, we know only as that which in all natural phenomena is continually passing from one portion of matter to another." From *Matter and Motion* (Prometheus Books: New York, 2002), 89. I find this to be a sober and accurate accounting of the state of physical knowledge, today as then (1876). Evidently, Maxwell regards our knowledge of both matter and energy as only indirect, mediated by their effects on our senses: we cannot claim to grasp what these things are in themselves. He simply accepts our ignorance, without trying to cover it up in positivist posturing.

¹⁰ See my paper, "Consciousness as Knowingness," in *The Monist* 91, no. 2, 2008.

question of what matter intrinsically is. Bodies are the divine made manifest. Is this a good theory?

No—but not because it is extravagant (or frankly theist). I set aside the question of whether such a notion of God deserves to be so called (Spinoza was suspected of being a closet atheist, of course). The germane question is whether pantheism of this stamp can serve the metaphysical purpose for which it was wheeled in. For the theory to stand a chance of working it needs to suppose that God is material in nature, since his nature is held to coincide with that of material bodies, which is to go against the more familiar idea that God is immaterial (hence the authorities' suspicion of Spinoza): God's parts, spread out across nature, must be material parts. But then we must have some conception of God's materiality; we must know what his materiality consists in. But that was our original question: what *is* it to be material? The concept of God gives us no further handle on the question.

The fact is that we have no conception of God's nature, as an immanent being, that would enable us to get purchase on our question about matter. We already have to know what matter is in order to form the idea that God is material. In the case of panpsychism at least we have an independent conception to work with—we already know what mind-stuff is—but in the case of pantheism we don't: not if God is to have a nature that includes the material. We can, if we like, simply *call* matter "God" (or God "matter"), but that is not to supply any *theory* about the nature of matter. We just have no idea what "God-stuff" might be such that it defines the nature of matter. I suspect, in fact, that the people who like to talk this way are tacitly assuming that God is *im*material ("spiritual"), so that matter turns out to be immaterial too (as "spiritual" as God). Really, deep down, they hold to a kind of panpsychism of the immaterial divine mind (whatever any of this may mean). Rocks and chairs and chamber pots are actually fragments of the Great Spirit that is God—and isn't that simply grand! But anyone whose aim is to keep matter material will not want to go this way, and will then be saddled with the circularity problem I have identified: what is the nature of the matter of which God is allegedly made? Either pantheism is a deified version of panpsychism or it is circular as an account of matter.[11]

So far I have been considering direct attacks on the problem, and failing to come up with a solution: we cannot complete the proposition "The nature of matter is...." Maybe we need to adopt a different approach, trying merely to locate matter within reality as a whole, so that we can see how it fits into the great scheme of things. Here is one attempt to do something along those lines. At one extreme we

---

[11] I suspect that the appeal of this type of doctrine, aside from the uplift it is felt to provide, is that "Spirit" is just the name of we-know-not-what, so that it fits perfectly our ignorance of the nature of matter: it doesn't remove the mystery, it advertises it—proclaims it. The word simply clothes the acknowledged epistemic mystery in mystical shivers. But whatever the objective nature of matter may be, I am sure it is not going to be fodder for mystics. Nor will it have the vagueness that the word "Spirit" trades upon. Matter, in itself, must be precise and tough-minded, not some sort of wishy-washy fairy-tale puffy stuff.

have nothingness—the state of there being nothing at all: zero being. Then we have empty space: not pure nothingness, to be sure, but pretty thin ontological gruel—just one notch above nothingness, we might say. Some thinkers, indeed, have doubted that empty space could be anything but nothingness—supposing that whatever reality space has is conferred by the objects within in (if we can even speak that way). For such thinkers, space is a kind of fiction we spread over the full-bodied reality of objects. Nothingness cannot be seen or touched and offers no resistance, but neither can space boast these positive attributes. It is being at its most minimal, just shy of zero.

Jumping to the other extreme, we have conscious subjects such as ourselves: rich and complex, pullulating with life, bursting with reality—high-wattage existence. Next down we have animals and plants, themselves many rungs up from mere empty space. In between we have the category of amorphous matter—brute spatial occupancy. It is a level up from space, but not yet at the heights of the organic (let alone conscious) world. It is the mere filling of space, at its most basic, with no further articulation—pockets of raw resistance. Metaphorically, these levels belong on a scale of ascending ontological *intensity*—more is packed in at successive levels. And the metaphor can be unpacked in terms of causal complexity: from zero causal powers to elementary ones to complex and sophisticated ones. The "definition" of matter is then this: it belongs low down the hierarchy of being, just above space—it is existence at its most lumpen and primitive (while not being entirely vacuous, like space).

I would not scoff at such ideas—they tell us something interesting about the ontological place of the concept of matter—but they do not really supply what we seek. All we can infer is that matter has a nature such that it occupies a certain niche in the larger scheme of reality, but we are not told what that nature is. We are still faced with the fact that while we know what the primary and secondary qualities of objects are, as well as the laws that govern them, we don't know what makes matter what it is: we still don't know what physical objects are ultimately made of. We know what matter does, and how it appears to us, and even its geometrical structure, but we are still ignorant about its inner nature. It occupies space, yes, but *with what?*

At this point defeatism might advertise itself as superior wisdom. Why can't we settle for saying that matter is made of *that stuff*? That is, granted we don't know—descriptively, discursively—the nature of matter, we can still advert to its nature, by means of indexical reference: we just stipulate that "matter" is to refer that stuff, whatever it is, that composes physical objects—or better, to avoid circularity, composes *these* (indexically identified) objects. Thus we can complete the proposition "Matter is..." by saying *"that stuff."* In other words, we select some paradigms and declare that matter is what these things have in common or what composes them all. On this view, there is no alternative to an ostensive specification of the nature of matter—and we should be content with that. After all, there is nothing to stop us from *referring* to matter, and hence forming thoughts *about* material things (and even about matter in general)—so let that suffice as our conception of matter.

However, such a view is surely faulty in principle, simply because there has to *be* something in common to all cases of matter, and its description cannot be *essentially* indexical. In standard natural kind cases we can always replace the initial indexical with a descriptive specification of intrinsic nature—as with "water is H2O" and the like—so it would be unprecedented to suppose that *in reality* only an indexical specification is possible. Imagine someone saying: "I know perfectly well what water is—it's *that* stuff," pointing at lakes, rain, and Evian. Contrast that person's state of knowledge with someone who knows that water is H2O. God surely knows the nature of matter non-indexically: he doesn't have to resort to paradigms and all the rest. The whole point is that while we can indeed refer to matter we don't know—descriptively and discursively—what the nature of what we are referring to is. The present proposal does nothing to gainsay that. The indexical proposal is therefore not an answer to our, perfectly legitimate, question; it is an admission that we can't answer it. I would say that the word "matter" means for us something like the description "the stuff that composes this and that," but we are unable to go any further in specifying what the nature of that stuff is; rather as someone might understand "water" to mean "the stuff that occurs in lakes, falls from the sky, etc" and be unable to determine that water is H2O. We are simply in the dark about the nature of matter.

So what is going on? It appears that our conceptual scheme bottoms out one step too early: we seem to have adequate concepts for the first two levels of the object, its primary and secondary qualities, including its physical laws, but then we are brought up short at the final level. (There is a question as to whether even our concepts of shape and motion are really adequate, but I won't press that point now.) Our ordinary concepts of physical objects don't encompass their full nature. The concept of matter that we have catches only what I have been calling its functional aspects—its causal and structural features—but it is silent on the constitutive aspects. If you study a textbook of theoretical physics, you will find that no serious definition of matter is ever given; it is assumed that you can extrapolate from some paradigm cases. There is much talk of particles and fields, but if you ask "particles of what?" no answer is forthcoming—except to say how particles behave. This is why there is room in physics for panpsychism to step in: nothing rules it out, because nothing is suggested in physics to fill the metaphysical slot it purports to fill. We can't say, "We know that panpsychism is false because physics has discovered that the intrinsic nature of matter is X, not mentality." And panpsychism has one major advantage: we at least know what consciousness is. How come we don't know what matter is? It seems to me that our basic predicament is this: *we are not acquainted with matter*. We are acquainted with its primary and secondary qualities (at least I am assuming this for the present), so that we have some claim to know what these qualities are, and we can be said to perceive physical objects themselves (*pace* Russell)—but we are not acquainted with what composes them. We experience colored expanses arrayed in some sort of space—primary qualities qualified by secondary qualities, apparently—but we don't experience the actual composition of the objects. Panpsychism entails this, since we don't experience matter *as* mental in nature (that is, when we are acquainted by perception with physical objects); and the ease with which we accept that consequence indicates that we do not suppose the essence of matter to be presented to us in the mode of acquaintance. We are acquainted with consciousness, so it is the kind of thing that would tell us the nature of matter, if that *were* its nature. But we are not acquainted with matter on the assumption that panpsychism is false. So when we cudgel our brains with the question: "What *is* matter?" we come up empty. We are (apparently) acquainted with the kinds of properties that characterize the microscopic realm, since these are just extensions of macroscopic properties—location, motion, attraction, etc—but we are not acquainted with the nature of matter at *any* level. Consequently, we simply don't know what it is. It is just like Kant's thing-in-itself.

My position here is essentially the same as Locke's, in his discussion of our "ideas of substance." In a famous passage he writes:

[I]f anyone will examine himself concerning his *notion of pure substance in general*, he will find he has no other idea of it at all, but only a supposition of he knows not what support of such qualities, which are capable of producing simple ideas in us; which qualities are commonly called accidents. If anyone should be asked, what is the subject wherein colour or weight inheres, he would have nothing to say, but the solid extended parts: and if he were demanded what is it, that the solidity and extension inhere in, he would not be in a much better case than the Indian before-mentioned, who, saying that the world was supported by a great elephant, was asked, what the elephant rested on? to which his answer was, a great tortoise: but being again pressed to know what gave support to the broad-backed tortoise, replied, something, he knew not what. And thus here, as in all other cases, where we use words without having clear and distinct ideas, we talk like children; who, being questioned, what such a thing is, which they know not, readily give this satisfactory answer, that it is *something*; which in truth signifies no more, when so used, either by children or men, but that they know not what; and that the thing they pretend to know, and talk of, is what they have no distinct idea of at all, and so are perfectly ignorant of it, and in the dark. (268)[12]

Thus we are like ignorant children when it comes to the real nature of material substance, however sophisticated we may be about the functional or structural aspects of matter. There is little more content to the idea than is contained in the word *something*. As he says, an adequate idea of matter is "remote from our conceptions, and

---

[12] Locke, *Essay Concerning Human Understanding* (Penguin Books: London, 1997), in "Of Our Complex Ideas of Substances." It should be observed that this claim of ignorance is quite compatible with Locke's earlier confident definition of matter as impenetrability in "Of Solidity." The latter definition fastens onto a *power* of matter, but it is silent on what might underlie the power—and the passage quoted in the text concerns the nature of the underlying substance. As he says, solidity "inheres" in matter, but what exactly it is that is thus inhered in is not known. We know that matter has the power to resist other matter, because we can see and feel its effects, but we have no idea of that in virtue of which the resistance takes place. We know what matter *does* but not what it *is*.

apprehensions" (270)—just not part of our conceptual scheme.¹³ Yet, as he also points out, that is no reason to doubt the existence of matter or to suppose that it lacks a robust objective nature; his claim, like mine, is purely epistemological.

I am not saying we *can't* know it: I wish to remain neutral on that question, and certainly see no proof that we are condemned to ignorance here—though our ignorance has the look of something very deep. We do seem to be confronted by a profound "mystery of nature," even if we might some day surmount the mystery. But I do think that as things are we lack acquaintance with matter, and therefore fail to have the kind of basis for knowledge that we have in the case of consciousness. This does not yet imply that we don't have an adequate *theoretical* concept of matter— one that transcends the fund of things with which we are primitively acquainted— but it makes it plausible that the lack of theoretical understanding traces to a basic cognitive deficit, possibly irremediable. We can't just put together prior bits of acquaintance to derive a complex concept for matter, because we don't have the *kind* of acquaintance necessary for such an exercise. Staring hard at physical objects will certainly not yield what we seek! The only concepts we have at our disposal are indexical and quantificational : "*that* stuff" and "*something* (we know not what")". Perhaps we can somehow come up with an adequate theoretical concept of matter, but it seems to me that acquaintance will not help us in that endeavor. At any rate, I don't think we have any such understanding now. We don't know what matter is, and so we cannot answer my title question. Nor do I have any idea how we might set about alleviating our ignorance (what kinds of experiments might decide the question?) This seems like an area of basic incomprehension—a conceptual blind spot.

The intellectual predicament I have outlined here strikes me as curious, even paradoxical. *Why* should we be so blind to the nature of matter? Matter is all around us, constantly perceived (at least *de re*), and we are *made* of the stuff, for goodness sake. Yet we seem not to have any inkling of what it consists in; there is remoteness in our references to matter—as if we only grasp its outer skin, its external form. Our knowledge of it contrasts sharply with our knowledge of mind. To be sure, we don't grasp *everything* about mind, but we surely grasp its essence—we are acquainted with what it *is* (just as we are arguably acquainted with color and number and ethical value).¹⁴ We know what the sensation of pain is, for example. We don't feel there is a giant blank at the heart of

---

¹³ In a passage I quoted in the Introduction, note 4, Newton declares us as ignorant of matter as we are of God: we know as little of the "inward substances" of things as we know of God's substance. Both Newton and Locke are sharply aware of how modestly our knowledge of the physical world extends: they have no tendency to try to squeeze reality into a box shaped by our own cognitive capacities. In short, we know something of matter, but we don't know everything (just as Newton says of our knowledge of God). Kant's noumenal reality is already deeply entrenched in these two thinkers (as also in Hume, on the proper interpretation). I think Newton had a very clear grasp of what he understood and what he did not understand. Subsequent philosophers have tended to celebrate him for what he knew, but have passed over his declarations of ignorance.

¹⁴ I could have mentioned space here, in contrast to matter: but though space may not be quite so hidden as matter, its real nature is still pretty puzzling (see chapter 3). What is true, I think, is that space *seems* more open to our comprehension than matter. We don't find the classical thinkers quite

concepts of conscious experience—as if we just don't know what we are talking about. But in the case of matter our concepts give out disconcertingly early, so that we feel that we are ignorant of the essential reality. We *ought* to know what matter is! Our ignorance here seems gratuitous, infuriating even (unlike, say, with the mind-body problem). And how come physics is such a spectacularly successful science, telling us so much about the physical world, and yet it fails even to begin to say what its ultimate subject matter really is? Its methodological operationalism seems insufficient to explain this lacuna—it is not that we could solve the problem of matter just by deciding to adopt a different set of methodological rules. The ignorance seems fundamental, and yet surprising. Something is blocking our access here—but what?

It is not surprising, then, that Eddington and Russell refuse to accept that we are thus benighted. For them, we *do* know the inner nature of matter: while physics (and unaided perception) tells us the outer properties of matter—what I have been calling its functional properties—introspection tells us of its inner being. When we are aware of our own conscious states, this *is* an awareness of the hidden dimension of matter—its underlying reality. All matter has this psychic dimension, but in our brains it becomes manifest to us through introspection. However bizarre this may sound, it answers a genuine hunger—to say what matter really is. Otherwise we find ourselves staring sightlessly into a total void, without even understanding how we come to be so blind. To the seeker after truth there is only one thing worse than intractable ignorance—*incomprehensible* intractable ignorance.[15]

ADDENDUM ON GRAVITY

The inverse square law tells us that gravity is proportional to mass and distance: the greater the mass the more the gravitational force, but the greater the distance the

---

so adamant about the mystery of space as they about the mystery of matter. We can see *into* space, after all, but we can't see into matter—so we think we have a better grasp of it. But that may not be so clear on deeper reflection—perhaps there is an illusion of understanding here.

[15] In the case of our puzzlement about the relation between brain and consciousness, we can make some rudimentary sense of why we are so baffled—because of the manifest incommensurability of the two realms (subjective versus objective, non-spatial versus spatial, intentional versus non-intentional, and so on). But in the case of our ignorance of matter no such explanation is forthcoming—it just seems brute and basic. We thus have no idea *why* we don't know what we don't know—no explanation of our evident ignorance. For people like Locke and Newton, however, the ignorance is only to be expected, because of the theistic viewpoint they bring to the world: for we can hardly expect to know what only God knows. Indeed, if we were to achieve such omniscience, the distinction between the divine God and human beings would be in danger of being erased. Thus our ignorance is bound up with our naturally inferior status in comparison with God. Locke and Newton don't need a special explanation for our lack of godlike knowledge, since they already know we are inferior to God. Humility is written into the order of things. But without that sort of theistic background, human ignorance of basic facts of the universe begins to seem badly in need of explanation—which is where evolutionary considerations are sometimes brought to bear. In any case, it is hard to find a feature of the problem of matter that makes our lack of comprehension itself comprehensible. It seems like a gratuitous lack.

less the gravitational force. Distance cancels the effects of mass, so that gravity dissipates with distance. But couldn't it have been invariant with respect to distance, not losing any of its power over longer distances? Light rays don't dissipate over distance, moving at a constant velocity through space and retaining their mass. Light reaches us with the same energy with which it left the sun, but the sun's gravitational field is much weaker at the earth's surface than it is closer to the sun. Similarly, is it *necessary* that gravity should increase with mass? Couldn't it have been constant with respect to mass, or even inversely proportional? The mass of an object is usually understood to be (or be measured by) its inertia—what it takes to move it from a state of rest or of uniform motion. So the law of gravity says, in effect, that the harder it is to move an object the more it makes other objects move; the more stuck in its ways it is the more it changes the ways of other objects. The sun makes the earth and the other planets revolve around it in a ceaseless dance of celestial agitation precisely because it cannot be easily moved itself; lack of mobility here begets extremities of motion elsewhere. That seems odd: wouldn't you expect that greater mobility in one object would lead to greater motion in others affected by it? Black holes are magnificently static in themselves, being of very large mass, yet they produce extremes of motion in objects that enter their gravitational field. The law of gravity thus hardly seems like an a priori truth, or even a very intuitive one. It seems odd and adventitious.

Or consider the primary qualities of gravitational bodies—say the shape and size of the earth. Gravity is completely independent of such qualities: it doesn't matter how extended over space a body is or what its shape may be, its gravity will depend solely upon its mass (as measured by inertia). Stars and planets tend to be spherical, but this has nothing to do with their primary physical power, viz. gravity; they could be triangular and their gravity would remain the same. But would we have been surprised and shocked to have discovered that gravity *does* depend upon such primary qualities—say, if it depended upon extension in space or on how far the body departs from being a perfect sphere? Isn't it more surprising that gravity wholly neglects such geometrical properties, so that how a body moves is indifferent to its distribution in space? Newton made the amazing discovery that of all the properties of bodies *only* mass determines gravity! This makes the world a lot simpler than we might have expected; the laws of motion turn out to be shockingly streamlined. Mass alone makes the world go round, though it is one physical property among many others.

The intuitive idea of mass is simply that of how much matter an object contains. This can be measured by inertia simply because the more matter there is in an object the harder it is to shift it—intuitively, it is *heavier*. So the law of gravity tells us that the more matter an object contains the greater will be its gravitational pull: if more matter *composes* it the greater will be its attractive force. In other words, the more matter you pack into an object the more immovable it becomes and the more it will cause movement in other objects. As I say, this seems far from self-evident: one wants to ask *why* things work this way, when they could easily have worked differently (compare the question of why light travels at the exact speed it does). What *is* it about matter that ensures that the more of it there is the more the gravity it exerts? What *accounts* for the law of gravity? To answer this question it would help to know what matter *is*; and that is putting it mildly—how can we hope to understand gravity *unless* we know what matter is? It is the nature of matter that gives rise to gravity, so to understand gravity we need to know the nature of matter. We need to know what it is *about* matter that guarantees that the more you have the more it pulls (and that the pulling dissipates over distance and is independent of other properties of objects).

The panpsychist has a theory of matter—it is mind-stuff. Panpsychism therefore entails that the more matter an object contains the more mind-stuff it contains—there is more consciousness in it. Either the object has lots of little minds in it, corresponding to its particulate structure, or these meld into one big mind in some emergent fashion. Thus the sun has more consciousness in it than the moon, say, which in turn has less than the earth. Then the law of gravity tells us, when conjoined with panpsychism, that the more consciousness an object contains the greater its gravitational pull (as well as the greater its inertia). But this doesn't *explain* the law of gravity; it simply introduces an even greater puzzle. For why should an increase in quantity of consciousness make a body attract other bodies with greater force? (We could ask a similar question about magnetic attraction: why should the presence of mentality make unlike poles attract and like poles repel?) So panpsychism does nothing to resolve the question about gravity I am raising (nor was it ever intended to, being proposed mainly as a way to make sense of the emergence of consciousness from matter).

I concluded above that we don't know what matter really is—its nature eludes us (and I have Locke, Newton, Maxwell, and others on my side). My conclusion, then, is that we don't know enough about matter to explain the law of gravity. If we knew what it is that fills space—what kind of stuff matter is—then we might be able to understand why more of this stuff generates greater gravitational attraction; but we don't have the former knowledge, so we are at a loss with the latter question. We know *that* something is so but we don't know *why* it is so. Nor do I think that translation into General Relativity substantially alters the picture: granted that matter deforms space in proportion to the quantity of it, the question still remains as to why this should be so. What is it about matter that curves space more the more matter there is? Matter *does* something to space, on this understanding, but we don't know *how* it does it—and the reason is that we don't know what matter intrinsically is. How can we hope to understand how matter relates to motion in space without having a clear conception of what constitutes matter?[16]

---

[16] Newton says at the conclusion of his great work: "I have not been able to discover the cause of those properties of gravity [such as its dependence on mass] from phenomena, and I frame no hypotheses": *Principia*, 442. He evidently regards his laws as omitting a crucial element in the

Perhaps, though, the connection to gravity might provide a way to think about the nature of matter. Whatever matter is it must explain the law of gravity; so its nature must be in some way bound up with the operation of gravity. Matter must be *such that* gravity arises from it. Whatever composes physical objects must be internally related to their capacity for gravitational effects. Matter is essentially a gravity-producing phenomenon. This must reflect on what it is in itself. But with these hints I must leave the topic, not having anything further to say.

explanation of gravity, which must reside in the hidden nature of matter. All the manifest characteristics of gravity, he assumes, must spring from this unknown cause. So Newton would agree that our lack of understanding of gravity has its source in our not knowing more about the nature of matter. He takes it that there must *be* some intelligible relationship here, but our ignorance of the cause of gravity prevents us from grasping it; he is *not* saying that his law is the end of the explanatory line—that there is no more to the world than events related by reliable laws. On the contrary, the laws rest upon real causes—which in this case we are ignorant of. We know that matter has the *power* of gravity, and we know how that power operates, but we don't know what the power proceeds from—though we can be sure it proceeds from something. The real cause of gravity is indecipherable to us.

3 The Possibility of Motion

Nothing could be clearer than motion. We see it with our eyes and we initiate it in our bodies. Surely we have a clear and distinct idea of motion, whatever may be the case with matter. We may not know all its laws and causes, but we are not blind as to its nature. We know what motion *is*. Don't we? Motion is naively conceived as the passage of objects through space over time. An object is at rest if it does not move from the place it is in, and it is in motion if it successively occupies distinct places. That is, motion is defined in reference to space. This commonsense idea is enshrined in the classic Newtonian conception of motion: space is an infinite and homogeneous medium, the parts of which are immovable, and the motion of bodies is relative to this static medium. Motion is thus conceived as "absolute," in the sense that whether an object is in motion depends solely on the relation between it and the unmoving medium of absolute space. An object could thus be in motion even in the total absence of other objects, so long as it occupied successive regions of static space.

If space itself could, *per impossibile*, be in motion, then the motion of an object might be invariant relative to a region of space, as that region moved with the object—so that we have motion without the occupation of distinct places. But granted the necessary immobility of space (i.e., the incoherence of the idea of space moving—for what would it be moving *through*?) the motion of objects is defined in relation to a space essentially at rest. The structure of space is unchanged by the objects within it and would remain what it is even if all matter were to be extinguished. Newton writes: "As the order of the parts of time is immutable, so also is the order of the parts of space. Suppose those parts to be moved out of their places, and they will be moved (if the expression may be allowed) out of themselves. For times and spaces are, as it were, the places as well of themselves as of all other things. All things are placed in time as to order of succession; and in space as to order of situation. It is from their essence or nature that they are places; and that the primary places of things should be moveable, is absurd. These are therefore the absolute places; and translations out of those places are the only absolute motions" (15).[1]

On this view, then, it is entirely coherent to suppose that the entire universe—all the matter there is—is moving at a certain fixed velocity through space, leaving one region of space behind as another is occupied. And that velocity could be arbitrarily large (short of the speed of light). Movement is simply a matter of a varying relation to a fixed and immutable space. "Absolute space, in its own nature, without regard to anything external, remains always similar and immoveable," Newton tells us (13). It is against this backdrop that motion occurs. The term "absolute motion" is therefore somewhat misleading, since motion is understood by Newton as relative to *something* (i.e., space), and it is not clear that it is meaningful to think of motion as relative to *nothing*.[2] But the space relative to which movement is defined is conceived as itself not in motion, necessarily so. Space is always and essentially at rest.

The Newtonian (and commonsense) conception is widely rejected in contemporary physics, as well as philosophy. Nowadays people tend to believe only in relative motion—the idea that objects can be said to be in motion or at rest only relative to other *objects*. Everything is at rest relative to itself, it is said, and everything moves relative to *some* other object. We are at rest relative to the earth, since we move with it, but the earth is in motion relative to the sun, so we move relative to the sun—though it also moves relative to us. It is entirely arbitrary what we choose to be at rest: a fly buzzing about the room is no more absolutely in motion than the whole galaxy is—we can in principle select the fly as our point of reference ("rest frame") and conceive everything else as moving relative to it. If the fly were the only object in the universe, it would make no sense to suppose it in motion, since motion is only well defined in relation to other objects. Objects thus replace space as the true coordinates of motion. It simply makes no *sense* to postulate an object moving at a uniform velocity through a space devoid of other objects. Motion, for the relativist, is just change of relative position, and this is entirely symmetrical: if you change position relative to me, then I change position relative to you. The spatial relations

---

[1] The quotation is from *Principia*, as are all other quotations from Newton, with page references in the text.

[2] Perhaps this is why the movement of space itself strikes us as immediately incoherent—because there would be no encompassing framework for it to be relative to. It cannot move relative to itself! Motion needs a medium distinct from the moving thing. Motion is necessarily relational: the question is what it is relational *to*—other objects or space itself. The idea of God moving poses grievous conceptual problems: he is not in space, so cannot move relative to space; nor can he move relative to ordinary objects. Is he then fixed rigidly in one place? (But he is not in a *place* anyway.) God is logically incapable of moving, like space (or numbers).

between us change, and that is the only reality that motion can have. There is no such thing as pure motion through space, or relative to it.[3]

The reason usually given to support the relative view is that it is impossible to *distinguish* rest from motion without the aid of other objects. Suppose you are in deep space and you see an object coming in your direction: it reaches you and passes by. You will suppose that it is in motion, while you are at rest. But if that other object is itself a person, she will have the same experience you had—of an object coming towards her and then passing by. She will take herself to be at rest and you to be in motion. Neither of you will have a sensation of moving yourself, since there is no air friction or any other reference body. If there is no other object around to settle the matter, then it is arbitrary to select one of you over the other as at rest: the right description of the case is just that you are at rest relative to yourself and she is at rest relative to herself—you both are moving relative to each other. It would be totally egocentric for either of you to declare yourself the still center of things, given the symmetry in your experiences. As Greene succinctly states what he calls the "Principle of Relativity," "Every constant-velocity observer is justified in claiming that he or she is at rest" (419).[4] From the point of view of other objects, we are moving; from our point of view we are not. That is all. There is no notion of absolutely moving or being absolutely at rest, since such a thing is entirely unempirical. Supposed absolute motion is completely undetectable.[5]

Two things need to be said about this argument: first, the premises are entirely persuasive; second, the conclusion doesn't follow from them. The premises are persuasive because our only *evidence* for motion comes from the observation of relative motion: absolute motion is, or would be, completely unobservable. Hence, as Greene says,

---

[3] See Nick Huggett, *Everywhere and Everywhen* (Oxford University Press: Oxford, 2010), chapter 9, for a clear discussion of different conceptions of space.

[4] Brian R. Greene, *The Elegant Universe* (Random House: New York, 1999).

[5] Maxwell puts the relativist position well: "Absolute space is conceived as remaining always similar to itself and immovable. The arrangement of the parts of space can no more be altered than the order of the portions of time. To conceive them to move from their places is to conceive a place to move away from itself. But as there is nothing to distinguish one portion of time from another except the different events which occur in them, so there is nothing to distinguish one part of space from another except its relation to the place of material bodies. We cannot describe the time of an event except by reference to some other event, or the place of a body except by reference to some other body. All our knowledge both of time and place is essentially relative. When a man has acquired the habit of putting words together, without troubling to form the thoughts which ought to correspond to them, it is easy for him to frame an antithesis between this relative knowledge and a so-called absolute knowledge, and to point out our ignorance of the absolute position of a point as an instance of the limitation of our faculties. Any one, however, who will try to imagine the state of a mind conscious of knowing the absolute position of a point will ever after be content with our relative knowledge." This is taken from *Matter and Motion* (Prometheus Books: New York, 2002; originally published 1876), 12. I sense a good deal of ambivalence in this passage, though it seems to come down finally on the side of relativism. There is a curious and obscure footnote appended to it, indicating Maxwell's ambivalence: "The position seems to be that our knowledge is relative, but needs definite space and time as a frame for its coherent expression."

every observer is *justified* in claiming to be at rest (assuming he or she is making no effort to move). We can observe the approach and recession of other objects, but we cannot observe mere passage through space without reference to perceptible objects. What goes for motion also goes for direction: direction can be judged only in relation to other objects. They approach from the front or the side or above; we cannot make a justified judgment of direction in relation to pure space. This is why we can seem to be at rest when we are not, relative to some other body, as with the rotation of the earth in relation to the sun: since we, standing still, are moving at the same speed as the earth, it *appears* that we are stationary. It is only when one object is moving faster than another that we can detect motion. If the universe really were moving through space all at once, as absolute motion permits, then we would be unable to detect such movement, since *differences* of motion are all we have to go on. Only relative motion is detectable, observable, and measurable. It is no argument against this to point out that we can feel motion by the pressure of air in our face, even when we can see no other object. The air is just another object moving relative to us, and we might as well declare that *it* is moving and we are stationary (this is why wind and locomotion feel the same). Relative to itself the air is at rest; relative to us it is moving. In empty space, rapid uniform motion feels just like stillness. You could not tell if you were moving or at rest unless other objects clued you in: and then it is arbitrary to declare you or those other objects as absolutely at rest. All *apparent* motion is therefore relative.

Newton was well aware of this epistemological point and indeed accepted it, so far as it goes. Immediately after introducing his preferred notions of absolute space and absolute motion, he writes:

But because the parts of space cannot be seen, or distinguished from one another by our senses, therefore in their stead we use sensible measures of them. For from the positions and distances of things from any body considered as immoveable, we define all places; and then with respect to such places, we estimate all motions, considering bodies as transferred from some of those places into others. And so, instead of absolute places and motions, we use relative ones; and that without any inconvenience in common affairs; but in philosophical disquisitions, we ought to abstract from our senses, and consider things themselves, distinct from what are only sensible measures of them. For it may be that there is no body really at rest, to which the places and motions of others may be referred. (15)

Thus he concedes that only relative motion is empirically significant, but he declines to conclude that there is no such thing as absolute motion. He distinguishes the sensible and measurable from the real and objective. That is to say, he asserts that the point about relative motion is merely an *epistemological* point. It says that we can only *know* relative motion, that only relative motion can be revealed to our sensory faculties (or any kind of measuring instrument or conceivable faculty of observation). There cannot then be apparent absolute motion—things appearing to move relative to empty space itself.

But it does not follow, he contends, that all *real* motion is relative, unless we assume a highly questionable verificationism—which is Newton's plaint in the passage quoted. It may well be that only relative motion is *detectable*, but it doesn't follow that only relative motion is *possible*. For there might be *un*detectable absolute motion objectively occurring in absolute space. The solitary object might be moving at a fixed rate through static space and yet no observer of that object could tell as much: it would *seem* stationary (suppose the observer to be moving at the same rate in the same direction as the observed object). But, of course, how things must *seem* is a totally different question from how things might *be*. There can be *illusions* of movement, after all, where something moving seems at rest and something at rest seems to be moving. From the fact that all apparent motion is relative it simply doesn't *follow* that all real motion is relative. There could, in principle, be verification-transcendent motion. We can't derive a metaphysical conclusion from epistemological premises—not without a heavy dose of unwarranted positivism. This is not yet to say that absolute space and motion are in the clear conceptually (see below), but it is to say that the customary rejection of these notions rests upon a dubious form of verificationism.[6]

To say that an object is moving undetectably relative to space itself is to say that it occupies distinct places at distinct times, irrespective of what space contains. This appears to be a well-defined and meaningful definition, whether or not such motion can be observed or measured; and it is surely our ordinary notion of motion. Objects move through the dimensions of space itself, leaving some regions to take up others. The relativist about motion must deny what seems an obvious truth: that the individuation of places is nonrelative. Suppose we have one object *a* moving relative to another object *b*. On our normal conception, one or both objects occupy distinct places at distinct times. But the relativist says that objects are at rest relative to themselves, even when they are moving relative to other objects. So from the point of view of *a* it occupies the same place over time, while from the point of view of *b* it (*a*) occupies distinct places. Are these places the same or different? Well, they are the

---

[6] Einstein expresses a startlingly candid commitment to verificationism in his discussion of simultaneity in *Relativity: The Special and General Theory* (Three Rivers Press: New York, 1961): "The concept [of simultaneity] does not exist for the physicist until he has the possibility of discovering whether or not it is fulfilled in an actual case. We thus require a definition of simultaneity such that this definition supplies us with the method by means of which, in the present case, he can decide by experiment whether or not both the lightning strikes occurred simultaneously. As long as this requirement is not satisfied, I allow myself to be deceived as a physicist (and of course the same applies if I am not a physicist), when I imagine that I am able to attach a meaning to the statement of simultaneity (I would ask the reader not to proceed farther until he is fully convinced on this point.)" (26). As an obedient reader of Einstein, I cannot in good conscience proceed farther, because I wholly reject verificationism. No doubt he would also endorse the same verificationist perspective on the nature of space. The most that can be concluded from Einstein's argument here is that a nonoperationalist understanding of simultaneity is of no relevance to physics as an empirical science—but that is quite compatible with allowing that it is perfectly meaningful and describes genuine facts in the world. Of course, Einstein was writing in the heyday of positivism, and so would not raise many eyebrows among his intended readers.

same from one object's point of view, but distinct from another's—and there is no objective absolute fact about which of these points of view gives us the real identity of the places involved. We have to say that place *p1* is the same as place *p2* relative to object *a*, but that *p1* is distinct from *p2* relative to object *b*. The identity of places is thus relative to the objects moving relatively to each other.

And not just *apparent* identity, *real* identity: whether one place is objectively identical to another depends wholly on the relative motion of objects within space. Whether *a* is in the same place over time relative to *b* depends upon whether *b* is moving relative to *a*. This means, of course, that totally empty space cannot have well-defined places within it—which is to say, it is logically impossible. To be sure, we don't *know* we have moved to a distinct place unless we can observe relative motion; but surely *that* we have moved depends entirely on whether the place we started out in is or is not identical to the place we end up in—and this is simply a fact about the structure of space itself. Places are individuated by their relations within the spatial manifold, Newton would maintain, and hence occupying specific places is a matter of the intrinsic geometry of space; it is not a matter of whether the objects in question are moving or at rest. To say that every object is at rest relative to itself, and that no other notion of motion is permissible, is to say that the place an object occupies is always identical relative to that object—but this commits us to supposing that places are the same or different only in relation to objects in relative motion.

In fact, we can give a formal proof, using Leibniz's law, showing that this position on the identity of places has to be wrong.[7] Suppose we take *a* to be at rest at *t* and to be at *p1*, and suppose too that *b* is moving relative to *a*. Then we can say, for *reductio*, that *a* is in the same place relative to itself over the time period *t-t'* but also that it is not in the same place over *t-t'* relative to *b*. So relative to *b* the place occupied by *a* after *t* is a distinct place *p2*. Accordingly, *p1* is identical to *p2* for *a*, but *p1* is not identical to *p2* for *b*. But surely *p1* is identical to *p1* for *b* as well as for *a*. If that is the case, then by Leibniz's law *p2* must be identical

---

[7] The proof mirrors David Wiggins' argument against Peter Geach's thesis of relative identity, according to which *a* can be the same *F* as *b* but not the same *G*. See Wiggins, *Sameness and Substance* (Oxford University Press: Oxford, 1980). The relativist about motion and space likewise thinks that whether two places are identical can depend upon an extra parameter: the places are identical with respect to an object considered at rest, but they are distinct with respect to another object relative to which the first object is moving—so the places are identical for *x* but not identical for *y*. Thus we have a thesis of relative identity. And the problem is that the place *x* is in at a given time must be self-identical, both for *x and y*, but that place can be referred to at a later time and produce a true identity statement for *x*, which means that *y* is committed to regarding that statement also as true, because of the substitutivity of co-denoting names according to Leibniz's law. The underlying problem is that the moving observer is committed to the truth of "the two places are identical for *x* but they are not identical for *y*, the moving observer," but how can this be, given that each place is identical for both *x* and *y*? It might be possible to wriggle out of this argument by some formal stipulations about the use of relativized identity predicates, but the force of the point remains: we don't want places to have their identity determined both by what is moving and what is at rest.

to *p1* for *b*, since *p1* is identical to *p2* for *a*. But then *p1* is both identical to *p2* for *b* and *not* identical to *p2* for *b*. As usual in these arguments, identity transmits all properties from one thing to another, and in this case the identity relative to *a* of places *p1* and *p2* carries over to *b*, so that these places must be the same relative to *b*, given that *p1* (or *p2*) is identical to itself for *b*. The argument works just by substitution of the terms "*p1*" and "*p2*" in statements of identity. What this formal argument brings out is the absurdity of supposing that a place might be identical to another place relative to one object (the one taken to be at rest) and yet this not be so relative to another object (the object moving relative to the first object).

A less formal argument can also be made, which trades upon Newtonian conceptions of motion and force. Consider the relative motions of the earth and sun. The earth moves relative to the sun, as it rotates on its axis (as well as in an elliptical orbit); but the sun also moves relative to the earth, as the sun tracks across the sky from dawn till dusk. According to the relativist, it is arbitrary which body we take to be at rest and which in motion. But this makes no sense of the force relations between the two bodies. The more massive sun forces the earth to move around it by way of gravity, but the earth does not force the sun to traverse the sky—it simply doesn't have enough mass to force any such thing. When a force is applied to a body, (accelerated) motion occurs (see Newton's laws), and forces vary with magnitudes such as mass. If we insist on regarding all motion as merely relative, we cannot recognize that some bodies exert more force than others. The earth has less gravitational force than the sun, so that the motion of the earth results from the gravitational force of the sun; but the earth cannot exert a like force on the sun, because of its lesser mass, so it cannot be causing the motion of the sun relative to itself. The sun moves as much relative to the earth as the earth moves relative to the sun, but these relative movements are differently related to the forces emanating from the two bodies. The sun causes the earth to move, but the earth does not cause the sun to move—yet the change of relative position is quite symmetrical. The asymmetry in the forces is not reflected in the symmetry of the relative motions. This is why Newton insisted on absolute motion: because of the link between motion and force. We don't want to say that the earth has just as much gravitational force as the sun *relative to itself*, simply because the earth can be stipulated to be at rest relative to the sun moving in relation to it. Force attributions can't be relative.[8]

---

[8] We might be able to gerrymander a relativized notion of force, whereby the earth has more gravitational force than the sun relative to itself, while the sun has more relative to itself. Then everything would exert massive gravitational force on all the things that move relative to it, thus possessing enormous relative mass; yet also that very thing will have tiny gravitational force, and hence tiny mass, compared to the things it is moving relative to. Mass turns out to be a relative quantity, turning on which objects we arbitrarily take to be in motion and which we don't. But this makes nonsense of the idea that some bodies exert more force than others according to their intrinsic properties, for example the quantity of matter in them (and can we really say that the earth has more matter in it than the sun relative to the earth, but of course less relative to the sun?) This is surely a species of nonsense.

It seems to me that these considerations about the individuation of places and the connection between force and motion strongly favor a nonrelative view of real motion (as opposed to apparent motion). Thus objects can be moving through space whether or not they are moving relative to each other; they just need to be occupying distinct places at distinct times. There is no argument against this commonsense view deriving from the epistemology of motion: the existence of nonrelative motion is compatible with accepting that all *apparent* motion is relative. Of course, physics is thoroughly operationalist in the way it defines its key notions, and hence believes only in what is measurable and detectable: but this is a philosophical standpoint not a piece of hard science. The influence on physics of pragmatism, positivism, and verificationism shows up in the way absolute motion is routinely disregarded; but we are not obliged to accept these philosophical doctrines. Even if absolute motion violates such doctrines, which it admittedly does, that is not a reason to declare the notion meaningless and irrelevant to physics. There may be real physical facts—such as absolute motion through space—that are not experimentally detectable. Not every physical reality has to be accessible to our faculties of observation and measurement. We cannot *equate* the real with the detectable—not if we are realists.

It may still be felt that absolute motion is suspiciously ad hoc and speculative. Why *should* motion through space be so undetectable—isn't this just a convenient way to introduce something metaphysically dubious? If there is such real absolute motion out there, shouldn't it show up somehow? If it is a *fact* about the universe, why should it be quite so invisible? The natural answer is that the unobservability of absolute motion actually follows from the very nature of space, as Newton indicates in the passage earlier quoted, and so is theoretically quite predictable. Space is uniform, homogeneous, the same throughout its extent: no matter where you are, it always looks and feels the same. It is *featureless*. It isn't like physical geography, where one stretch of land looks quite unlike another, so that you can tell where you are by observation: space is monotonously the same everywhere. Therefore, you cannot tell your position just by observing the way space is there: it appears the same no matter where you are.

Imagine that space had different colors in different regions of it; then you could tell you had moved by observing the change of color. But space is colorless and uniform, so you can't tell. Because one place is just like another, it is impossible to tell whether you are moving through space or not just by consulting space. To tell whether you have moved you need to consult the varying landscape of physical objects (and then you get only relative motion), but motion through space itself presents no variation of landscape. Absolute motion is unobservable simply because all parts of space are inherently indistinguishable; they are all "qualitatively identical." The passing spatial scene *as such* is invariant in its appearance. Because of the very *nature* of space, then, rest and motion present the same appearance. The epistemological inaccessibility of absolute motion is therefore predictable from the metaphysics of space. Space is *built* not to reveal movement through it. Our only way to tell that we are moving is to observe the relative movements of other things, but this is an indirect and fallible way to judge motion through space itself—there being no direct way available in the nature of the case.[9]

Motion is therefore a fertile area for skeptical hypotheses. What is really moving and what is not? We know that *some* things must be moving through space, because we can observe relative motion, but how can we establish which? Maybe the whole universe is moving at a thousand miles a second through space. Maybe the only thing not moving is that fly buzzing about I mentioned earlier—everything else changes place but not that little fly! Galaxies are moving away from each other, propelled by the force of the big bang, but we cannot judge whether any of them might in fact be at rest in space. Maybe the earth really is the only stationary object in the universe, after all—it alone never leaves its absolute place in space. It moves around the sun, but that is just relative motion, and doesn't settle which thing is really moving through space itself.[10] Of course, in the ordinary sense, we have reason to believe that objects are in nonrelative motion and are able to gauge their speed, since relative motion gives us (fallible) evidence of absolute motion: but the skeptic would be within his rights to point out the deep epistemic gap between relative and absolute motion. To common sense, the branches swaying before me really are moving through space, changing their absolute location; but the skeptic can argue that this *might* be an illusion, since everything *else* might be moving instead. This may be farfetched, but it is not logically impossible. We are certainly familiar with cases in which we think something is moving and it is not—and who can say how extreme such errors may be? Still, the possibility of skepticism does not generally make us reject our commonsense views—say, about the external world and other minds—so it is not clear that it should disturb us unduly when it comes to absolute space. Isn't this just one more area in which evidence and fact fail to converge perfectly? There is absolute motion, but we are fallible in our judgments about when it is occurring.

[9] Suppose you are driving down a very monotonous highway in a featureless desert. There is no way to distinguish one place from the next; it all looks the same. Is that a good reason to deny that the desert and the highway exist? Not knowing whether you have come full circle to the same place, as opposed to reaching a qualitatively identical place, is insufficient to declare that you are not traveling through the desert: you might be or you might not be—it's just hard to tell. Space for the Newtonian is like that desert highway: the difficulty of determining whether you have changed location should not be converted into a proof that absolute space does not exist—it might merely reflect the inherently featureless character of space. Just because something is monotonous is no reason to deny its existence.

[10] We can infer absolute motion in the earth by noting that the sun is exerting a motive force on it in virtue of its mass—where there is force there must be real motion. But the mere observation of apparent motion is never by itself conclusive evidence for real motion. The earth *might* be stationary, despite the sun's exerting gravitational force on it, because some other body is exerting a counter-force. It always seems possible for a skeptic to come up with a scenario that upsets our usual ideas about what is really moving. Apparent motion never *entails* real motion, because it is always logically possible that the appearance of motion results merely from a change of relative position, which is compatible with the object in question being at rest absolutely.

Should we conclude that absolute motion is conceptually in the clear, while relative motion is in serious trouble? Well, we have so far found no persuasive argument against absolute motion, just some bad old positivist arguments, and some reason to look on it with favor. But there might be other considerations that weigh against absolute motion, and certainly the entrenched historical hostility to it suggests that the case may be more problematic than we have hitherto recognized.[11] Maybe the bad verificationist arguments spring from something deeper and more troubling. That is what I now intend to argue: absolute motion is not as clear as it may initially appear, even to the resolutely anti-verificationist. Space, as Newton conceived it, echoing common sense, is infinite, boundless, and homogeneous. It could exist without any material object within it. Places composing this endless manifold are individuated by the structure of space itself. A single object could be introduced into this space and begin moving in a determinate path—with a certain speed, in a certain direction, carving a certain shape of trajectory, accelerating and decelerating. All that is required is that places within space should be successively occupied and a particular form of movement is thereby determined. But is this really credible? Are we perhaps smuggling in some nonspatial reference point and tacitly taking the alleged motion to be relative to that object? I suspect that one of the things we do in thinking about such solitary movement through space is to bring in the idea of a boundary or border to space, because of the difficulty we have in conceiving of infinite boundless space. We sneak in the idea of a limit to space's extent—an end to space. Then we think of the movement of the solitary object as *approaching* such a boundary—getting closer to it. But that is to think in terms of relative motion. What we are supposed to be envisaging is motion through a medium in which the moving object never gets nearer to or further away from *any* object; in infinite absolute space motion loses such a relational property. The object never gets nearer to the edge of space, because space has no edge. Yet we irresistibly think in such terms, as if space is a kind of enormous box, because infinite space is so hard to comprehend. The moving object will never get one jot nearer to the end of space, no matter how fast it goes and for how long. Does it now still seem clear to you that it is really *moving*? It is certainly not moving *towards* anything. What direction is the object moving in? None that can be specified in terms of physical coordinates: so does it *have* a direction? What shape of trajectory does it carve—circular or zigzag, say? None that can be defined by reference to another object: so how can it *have* one or the other? There

---

[11] The rhetoric against absolute motion always seems to outrun the strength of the standard arguments against it, suggesting that the official arguments are not the root of the real objection. Poincare, in *Science and Hypothesis* (Dover Publications: New York, 1952; originally published 1905), recommends relative motion for two reasons: "the commonest experiment confirms it; the consideration of the contrary hypothesis is singularly repugnant to the mind" (111). The repugnancy of absolute motion surely cannot stem merely from the fact that it is empirically undetectable (like an inverted spectrum case); it seems to derive from the very idea of absolute motion as a logical possibility—that is, from internal incoherence. But what is this incoherence supposed to be exactly?

is supposed to be a definite form of movement on the part of the solitary object, but it seems completely unanchored, a kind of fictitious imposition. Be careful, now, in assessing the claim that you do not introduce yourself into the picture, in addition to that nonexistent border: don't imagine yourself in the space observing the movement of the solitary object. For if you do, then you are illicitly taking your own body as a reference point for the movement of the object. Remember, the object must have a definite type of motion quite independently of anything other than homogeneous space itself. But what, one wants to ask, could this possibly consist in? What is the *difference* between an object moving at a million miles an hour in a figure eight and an object moving at two miles an hour in a straight line? The Newtonian will reply that different paths through pure space are carved: but this begins to sound empty, pending some account of what makes the regions of space distinct from each other. It is much like the question of size: what could constitute an object as having a determinate size in an infinite space containing only that object? Taking up more space, the Newtonian insouciantly replies: but do we really have a handle on what this means, granted the boundlessness of space and the lack of any object with which to compare the lone object? The case seems different with shape: we can say what it would take for an object to be triangular rather than rectangular in absolute space—because we can talk of angles and sides. But with motion we just have the blank affirmation of determinate movement, wholly undetectably, without anything further we can say about it. There is nothing for such movement to *consist in*. It is not merely that we cannot *verify* the characteristics of the solitary object's path, though we clearly cannot; we don't really have a clear and distinct idea of what such a path would *be*. The reason we don't see this point at first blush is that, for understandable reasons, we tend to smuggle in tacit reference points when we try to imagine the motion in question—borders to space or ourselves as observers. But these must be rigorously excluded if we are to assess the coherence of pure absolute motion.

Time is very similar. Can we really make sense of the duration of an event in absolute time in which only that event occurs? Granted that *judgments* of duration are necessarily relative to other events, is it clear that the *fact* of duration objectively exists in a time series containing only a single event? Here again we must be careful not to smuggle in a reference point, such as our own conscious events as observers—so that we judge a given event to be, say, twice as long in duration as some conscious event in our minds. In absolute time, events are supposed to have determinate duration independently of anything but the bare passage of time itself: they have "bare duration" rather as objects in absolute space have "bare motion."

But don't we need a coordinate system, based on a choice of reference point, in order for duration and motion to make sense? The idea of free-floating motion looks like a trick of the imagination—a possibility that isn't really one. Certainly, our imagination is easily tricked into envisaging such pseudopossibilities—modal thinking is notoriously fallible. It is easy to think you can imagine a disembodied mind, for example, but more careful reflection can cure you of such excesses: impossibilities are not always easily recognizable as such.¹² Perhaps the notion of absolute motion is like that: we think it is possible until we really think it through, excluding all illicit intrusions. Consider a whole series of empty universes save for the solitary moving object in each: can we really understand the idea that vastly different patterns of movement characterize these objects, even though observably they all look exactly alike and there is no causal difference between these alleged motions?¹³ Indeed, can a solitary object be said to be in one place rather than another in a universe of infinite and homogeneous space with no coordinate system supplied by other objects? (Much the same question can be asked about being at one time rather than another in otherwise eventless time.) The idea of the motion of an object in an otherwise empty space would seem to be a prime candidate for a claim of *indeterminacy*—there is just no fact of the matter about what its particular pattern of motion might be. To think it has one is to fall victim to a trick of the imagination, abetted by illicit reference objects.

If we could believe in an ether, we might have a basis for defining absolute motion: the object is moving relative to the quasi-material ether, conceived as pervading all of space. But that is really no help in giving content to the concept of absolute motion, because the ether is another one of those illicit tacit reference points that smuggle in relative motion. The object carves out a path through the ether, which exists over and above pure space, in either a continuous or particulate form: but then the object is moving relative to this ethereal thing, which may or may not itself be moving. The ether is no better than an all-pervasive invisible gas in grounding absolute motion. In addition, of course, there is no ether.

Things might be different if points of space were intrinsically labeled in some way, so that their positions in infinite space were written into them. In the case of an infinite sequence such as the number series, each number has a determinate position in the infinite totality: each point is located relative to the other points in a determinate way—2 comes before 6, say. But infinite space is not like that, because the places are homogeneous, unlike numbers: the points or regions of space are all alike in intrinsic nature, unlike numbers. So nothing in the nature of a particular place fixes its

---

¹² Such as the alleged possibility of zombies that duplicate the body and brain of a conscious being, or water not being H2O, or people having different origins, or wood tables being made of ice. These all seem possible *prima facie*, but careful reflection can convince you that they are not really possible. The human mind seems constitutionally gullible about possibility, perhaps because of our overactive imagination: we are modally credulous by nature, all too ready to think anything is possible.

¹³ Be careful not to introduce counterfactuals about objects into the picture, as in: if the universe with a solitary object in it *were* to have another object introduced into it, then the relative motion would be thus and so. We imagine another object in there as well in order to give content to the idea that determinate motion through space is occurring. For that, again, is to smuggle in relative motion, i.e., something that functions as an axis of motion. The fact that the new object is merely possible or imaginary doesn't absolve it of illicitness, because we are still trying to conceive of absolute motion through the crutch of relative motion.

location in the totality—only its relation to other places. But nothing ties down these other places—certainly not a boundary. Nor can we appeal to God to save absolute motion, holding that motion is all relative to him as omniscient observer: for, again, this is to introduce an illicit reference point over and above space itself. If God is in space with the solitary object, it is not solitary after all; but if he is outside space, how can he operate as a coordinate system? I strongly suspect that the *prima facie* intelligibility of absolute motion depends upon an imaginative mixture of the godlike and the human: we envisage ourselves as occupying a godlike position as observers of the universe, and then we imagine motion occurring relative to ourselves as so imagined—we think of the object as accelerating away from us, say. But this is, to repeat, an illegitimate procedure: we were supposed to be conceiving of motion as occurring independently of *any* nonspatial reference point. You have to try to imagine determinate motion in a universe with no conscious observer, human or divine, no ether, no spatial border, and no other object: I submit that you will find that this is none too easy a thing to do. You have simply subtracted away too much from the ordinary idea of motion for it to survive in such reduced circumstances. All our notions of *direction, towards, up* and *down, approaching, receding*, and so on, depend upon a coordinate system constructed around a reference point—and without such a system they fail to make proper sense. But independently of such notions the concept of motion becomes empty.[14]

Where does this leave us? Nowhere very comfortable, I fear. Relative motion is not real motion, and anyway runs into severe problems. Change of relative position is fine, but does not furnish an account of *motion* as such. It is entirely noncontradic-

---

[14] I may be a staunch antiverificationist, but even I am not willing to declare that something must make sense precisely because it is *un*verifiable. I don't hold that what is unverifiable is thereby meaningful (or that what is verifiable is *ipso facto* meaningless!). But I need extra arguments to move me to question the meaningfulness of a notion. The case of disembodiment seems very like the case of absolute space: we distinguish between mind and body, and between space and objects, and then we think we can strip away one side of these distinctions and leave the other side perfectly intact. We take the body from the mind or the objects from space—and we entertain the notion that nothing inherently changes in mind or space as a result of these subtractions. But the subtracted items might well be necessary conditions for the supposed remaining items, especially when it comes to questions of individuation: what makes *this* mind what it is once the links to the body have been severed, and what makes *this* place the place it is once a coordinate system has been eliminated? Or, to go one step further, there are the alleged possibilities of the bare particular and the transcendent ego: can you strip away all the properties from a particular and have a determinate thing left, and can you really conceive of yourself existing without any physical or mental characteristics? Surely, in these cases, too much has been stripped away for the remainder to count as a particular or a self; to suppose otherwise is a trick of language or the imagination. I actually used to be a supporter of absolute space, with solitary motion in empty universes, until I became convinced that I had not stripped away all that I needed to in order to have a pure case of absolute motion—and then it struck me forcibly that what I had left was too indeterminate to add up to real motion. I hadn't ruthlessly eliminated the illicit reference points from my thought; rather as someone might convince themselves that total disembodiment makes sense without really ridding their thoughts of all remnants of body—as it might be a ghostly or gaseous body.

tory to allow that an object changes its position relative to another object and yet *does not move* (because the other object did). On the other hand, absolute motion seems not to be finally intelligible. Change of relative position is not motion, though it is a perfectly coherent concept; but absolute motion is not coherent, though it *would* be real motion if there were such a thing. Neither of the traditional theories therefore gives a satisfactory account of what motion consists in. Yet these seem to exhaust the available conceptions of what motion might be.

What are our remaining options? One would be to declare motion unreal, following zealous Zeno, but for different reasons: nothing moves. Another would be to go idealist about motion: what we mean by "motion" is just a propensity to produce sensations of motion in us—treat motion as a secondary quality, in effect. There would then be no objective motion, just our sensory impressions of it: a universe without observers would be a universe without motion (as with subjectivist accounts of color). There is motion, but it reduces to our impressions of motion—we spread it on a world that does not independently contain it. But there is a third option to consider: the word "motion" stands for something we know not what. We tried to give content to the word by appeal to change of relative position or the supposition of absolute space, but these attempts failed. We appear to have no further alternatives available as descriptive analysis.

But this does not mean that the word fails to denote something real: it means merely that we have no positive conception of what that thing is. In other words, we can *go structuralist* about motion. We thought we knew what motion is, as we may have thought we knew what matter is, but we were wrong—we don't know what it is. The word "motion" is just a label for an unknown quantity that we know only abstractly and by its effects on our senses; we don't know its intrinsic nature. It is easy, even for a committed structuralist, to suppose that motion is transparent, though other concepts of physics are not. Since motion is the *explanandum* in physics, we tend to assume that we grasp its essence, even if we fail to grasp the inner nature of the *explanans*: but that last refuge from structuralism now seems in danger of extinction. Motion is not the *datum* we had assumed. Whatever motion is, it seems remote from our usual conceptions of it, confused and erroneous as these are. We grasp it only abstractly, by means of mathematical representations, i.e., structurally.[15]

---

[15] In addition, we grasp the effects of motion on our minds—as with visual experiences of motion. But now we see that these sensations do not reveal the objective nature of motion; they are merely symptoms or signs of something we-know-not-what. The case is like our sensations of electric current: these sensations are effects of electricity, but presumably no one will want to say that they reveal the objective essence of electricity—any more than sensations of heat reveal the objective essence of heat. (Maybe sensations of color reveal the essence of color.) When I have a visual experience of a bird flying by, this is not a direct revelation of the nature of objective motion, but merely a mental symptom of whatever physical fact is causing my sensations (for one thing, the perceived motion is relative, but whatever real motion is, it is not relative). As the old philosophers used to say, there is no *resemblance* between my sensation and objective motion: I do not experience motion as it is in itself—any more than I experience heat, electricity, gravity, and matter as they are in themselves. As Russell would put it,

Why does this position seem counterintuitive and hard to accept? I think it is because of *vision*: in seeing things we seem to be presented with motion in a very direct and transparent fashion, and this forms the basis of our conception of what motion must be. The other senses don't present motion anywhere near so forcefully—so that here we feel that motion is something *inferred*, not given. A physicist without vision might find it much easier to accept that motion is known only abstractly and not in its intrinsic nature. Vision gives us an *impression* of understanding, but it turns out that we have no intellectual resources with which to form an adequate conception of motion. Motion as it is in itself does not coincide with motion as-we-perceive-it (Special Relativity might already have alerted us to this possibility). The word "motion" is a theoretical term, denoting an entity or process that we know about only remotely; it is nothing like "pain" or "red." Motion is as unknown as matter.[16]

This result is actually quite philosophically welcome to the hardened structuralist. Before, we might have supposed that physics divides into a transparent part and an opaque part, with motion falling into the transparent part and matter falling into the opaque part. That seems worryingly fragmented: why should motion be so transparent if what moves (matter) is not transparent? But now we see that *both* are opaque: thus we achieve a theoretical unity in the basic terms of the science. Motion, considered objectively, is as unclear as everything else; it is merely that we surround this unclearness with plausible-sounding would-be explanations that don't in the end stand up to scrutiny—relative motion and absolute motion, notably. The fact is that in physics we represent motion mathematically, and this works well enough for physics to proceed. In physics we describe motion by assigning numbers to space and time—units of distance and duration. Thus we say that a projectile is moving at, say, twenty miles an hour: it covers twenty units of distance (miles) in one unit of time (an hour). Because we can describe space and time mathematically, we can describe motion mathematically—and that description is what physics needs and can work with.[17] We don't need to grasp what motion is in itself—which is fortunate,

I am not acquainted with motion, as it objectively exists: I do not *see into* its essence. Newton never supposed otherwise: his absolute motion was an intellectual construct, a product of abstract reason, not a perceptual given—indeed, he thought that only change of relative position is perceived by the senses. The structuralist agrees with Newton on the imperceptibility of true motion, but adds that his abstract absolute conception fails to fill the cognitive gap: we have no positive accurate conception of objective motion, but only a mathematical description backed by laws, and supplemented by sensory symptoms. Motion, for the physicist, is just one set of mathematical units plotted against another—units of something called "time" plotted against units of something called "distance." But what are time and distance? They are simply, for the physicist, that which is measured by such scales as hours and seconds and feet and inches. Our conception of them is purely scale-based.

[16] On the unknown character of matter, see chapter 2, and elsewhere in this book.

[17] I think the mathematical character of physics arises from a very straightforward fact, namely that time and space are both susceptible to simple mathematical description, by means of scales of measurement. This makes motion itself susceptible to mathematical description, since it is straightforwardly definable in terms of time and space: how many units of space are traversed in how many units of time. So motion can be smoothly mathematized. But motion is the central concern of

because we don't *have* an adequate grasp of it in itself. We have *in*adequate ideas of motion, in the form of relative motion and absolute motion, and messy confused mixtures of the two. Motion, in physics, is just the name we give to whatever it is in reality that fulfils the mathematical description we assign to it—units of space and units of time. Motion is just whatever is so measured.

I now want to make some remarks about the implications of the foregoing discussion for Special Relativity, specifically the speed of light. We are assured that the speed of light is always measured as a constant, no matter what the motion of the observer. If you are moving in the same direction as a light beam (or away from it), you will always measure the speed of light as c; you can never "catch up" with light, even a little, no matter how fast you travel. This is a very remarkable property of light (and very hard to credit), which led to Einstein's overturning of traditional ideas about space and time. It means that no matter how fast an object is moving, the speed of light away from it will always be c. This is generally taken to show that the speed of light is a constant—that it can neither speed up nor slow down (in a vacuum: it slows down in media like water and glass). But what we really should say, more exactly, is that the *relative* speed of light is a constant: light is always moving at the same speed relative to other objects moving through space. The laws of physics, we are told, remain the same in any rest frame, so light must travel at c whichever rest frame is chosen—even when that frame is moving relative to others.

Now let us accept that orthodox story (and we should not underestimate how hard it is to do that). Then, if we were to accept Newton's notion of absolute motion, as a backdrop to relative motion, the right conclusion to draw must be that the *absolute speed of light varies*.[18] The only way for light speed to be constant relative to a

---

physics—its prime *explanandum*. Since motion can be represented quantitatively, by means of scales of space and time, anything introduced to explain it inherits that quantitative character: forces will now be measurable in the same way, in terms of the motion they produce, which in turn cashes out in units of time and distance. Thus the mathematics works back from space and time, to motion, to the causes of motion, to the laws of motion. Physics is mathematical because space and time are—that is the *sine qua non* of mathematical physics. If it were not possible to apply measurement scales to space and time (particularly ratio scales), physics as we know it would not exist: for then motion could not be measured in terms of units of space and time, and consequently the forces that produce motion would have no mathematical structure. The precision afforded by scales of measurement for space and time lies behind the precision of physical laws. But none of this requires that we have any real insight into the intrinsic nature of what is thus measured. The terms "space" and "time" could be just uninterpreted place-holders and yet we could still state the laws of physics—"An X takes up n units of Y in order to be measured by m units of Z" or "An electron takes up n units of time in order to be measured by m units of distance." In practice, of course, we rely on our commonsense notions of space and time to read meaning into physical laws, inadequate as these may be; but we have no substantive account of what these terms refer to. We merely have mathematical structure combined with some vague and dubious intuitive ideas.

[18] It is perfectly consistent to hold that absolute and relative motion can co-exist, so that the constancy of the relative speed of light can occur against a background of varying absolute speed. This is a coherent world, whether or not it is the actual world. At any given moment light might have two speeds: its invariant relative speed, and its varying absolute speed. The measured constancy of its relative speed does not logically rule out a varying absolute speed.

moving observer is if it can move at varying speeds relative to space. Suppose I accelerate in the direction of a light beam, and suppose that its speed relative to me remains constant: that can only be so (on the assumption of absolute space and motion) because it too accelerates—not relative to me, but relative to space. I am moving absolutely through space with a certain velocity, and the light beam maintains its speed relative to me; so it must be going faster through space than it would be if I were moving more slowly. In other words, if motion is absolute, the invariance of the speed of light relative to a moving observer entails that the absolute speed of light varies. If relative speed is constant, absolute speed must be variable, given that the observer can be moving too.

That is not what Einstein intended—he had rejected absolute motion on broadly positivist grounds (following Mach). The result is certainly paradoxical, implying as it does that light can accelerate through space without any force being applied to it—as well as violating the spirit of Maxwell's theory of electromagnetism. A constant absolute velocity for light is what we would expect (and hence a variable relative velocity), but that turns out to be wrong if the relative velocity is actually constant. We could try to question that principle, but apparently it is experimentally well confirmed. The only way out of the paradox that I can see, if absolute motion is taken to have any validity at all, is to suppose that the absolute speed of light *is* constant, which means that observers can reduce the difference of speed between themselves and light, hence varying relative speed—but that relative speed is always *measured* as constant. It is thus not really constant, but only apparently constant. From the point of view of the observer, the speed of light appears as a relative constant, but it is not constant *sub specie aeternitatis*. The apparent constancy is a kind of *interaction* effect—an effect of the measurement relation between light and the observer. We can of course deny absolute motion to escape the paradox, so that light never has to speed up in absolute space in order to maintain its constant relative speed: but for anyone sympathetic to some notion of absolute motion, we must find another way out.

The point I am making is that we are not *compelled* to reject absolute motion by the facts of physics, though we do need to interpret those facts—particularly, the measured constancy of light speed relative to any observer—in the right way. The alternative interpretation is that the absolutely constant speed of light is measured in such a way that it appears to remain constant relative to a moving observer, while not actually doing so. That is, it is measured as constant relatively (and this is an objective fact about it) even when it is *not* constant relatively. It moves through space always at the same velocity, so that an observer can reduce the relative velocity between himself and it, but it is always *measured* as c relative to the observer. Why this should be I don't know, but the logic of the situation allows it as a possibility; and I certainly find this interpretation of the data much more satisfactory than the orthodox position that you can never reduce the difference of speed between light and any other moving object. In sum: light moves relative to space just like anything else, and it has a constant absolute speed. Other objects also move through space, varying their speed, some faster than others. The relative differences of speed between light and other objects must accordingly vary with their speed and direction of movement. That this is not measured to be so reflects the process of measurement, not the big metaphysical picture.

That, at least, is the kind of thing a Newtonian would say about the constant speed of light, and I think it should be taken seriously. But didn't I argue earlier that absolute motion is a dubious concept? Yes, but I didn't argue that relative motion is the only kind there is—on the contrary. I think that genuine motion is nonrelative in nature—I just don't have a positive conception of what this kind of motion might be. Objective real motion is an I-know-not-what. And this applies to the motion of light as much as to any other motion. The motion of light is certainly extremely puzzling, given the empirical fact of its constant relative velocity. Given what I said earlier, we can now note that light has a type of nonrelative motion that underlies the relative kind that we observe. Does this vary as objects accelerate towards a light beam? Again, that seems indicated if we take the constancy of relative speed at face value. But do we have to take it at face value? Perhaps, again, we measure relative speed as constant, but it is not so in reality. The light beam is moving nonrelatively, in some way we do not comprehend, and it does so unvaryingly: but when we observe its relative movement, which is a perceptible type of movement, we find a constant—which is best explained as a superficial measurement effect.

The general point I want to make here is that the paradoxical nature of light should be understood in the context of our lack of grasp of the nature of objective motion. We don't know what real motion is, so we should be careful making bald claims about how light must be moving—specifically, that its motion is entirely relative and yet a constant. Newton was right to reject relative motion as real motion, even if his positive theory is hard to make sense of; and so we cannot rest content with an account of the motion of light that works only with relative motion. My suspicion is that what we call the movement of light is so far removed from our ordinary conceptions, and so specific to light itself, that the orthodox story is a travesty of the underlying reality. The observed constancy of the relative speed of light, paradoxical as it is, is a clue to how odd (relative to our conceptions) the objective motion of light must be. What is the motion of light such that such a thing is possible? Here, again, we should not rule out a verdict of deep mystery: if we don't know the real nature of the movement of a simple stone, how much less do we grasp the real nature of the movement of light? Do these things even move in the same *sense*?[19]

---

[19] I am acutely conscious in these remarks of going way out on a limb, and treading on the toes of orthodox thinking. I am really trying to open the subject up a bit by outlining some theoretical options. I am also conscious of how difficult and obscure the topic is.

Physics is the science of matter, supposedly. But it is striking how much of it is concerned exclusively with motion.[20] From a metaphysical point of view, this must seem odd, since motion is a contingent property of matter: we can easily imagine worlds in which space-occupying objects don't move—is there, then, no physics in such a world? The speed at which objects (including light) move is a contingent property of them—the same objects could move at a different speeds—and that they should have *any* speed seems likewise contingent. The universe could exist in a "steady state," as a matter of metaphysical possibility: nothing moving, ever. This might be because the matter is subject to no forces, or because the forces balance out. But couldn't matter exist, in some form, and yet not be pushed around by any forces? Certainly, it could be subject to different forces, or forces with different properties (gravity that did not obey the inverse square law, say). Physics, as we have it, is basically concerned with extended solid things that are subject to forces that generate movement, but the connection between body and force seems contingent: nothing in the idea of body seems to entail the forced motion of body, despite the salience of motion in our universe. To occupy space is not the same as moving through it, absolutely or relatively. This, presumably, is why Descartes and Locke did not count motion as the essence of body: motion is, strictly, *extrinsic* to body.

Then why is physics so obsessed with motion? That it is so obsessed is undeniable. All the four basic forces recognized in physics—gravity, electromagnetism, the strong force, and the weak force—all concern motion, even when (as with the strong force holding the constituents of the atomic nucleus together) the force works to inhibit motion (the strong force explains why the constituents don't fly apart). The standard forces are all about attraction and repulsion—objects moving through space. That is not a conceptual truth about the notion of force: we can apparently conceive of a force that changes the shape or color of objects but does not alter their location.[21] A force is just what makes things happen—but in physics the happenings all seem to involve motion. The two main forces are even *defined* in terms of motion, pretty much—gravity is that which is responsible for a certain class of movements, terrestrial and celestial; electromagnetism produces charge, which is a power

---

[20] In *The Problems of Philosophy* (Oxford University Press: Oxford, 1968; originally published 1912), Russell writes: "Physical science, more or less unconsciously, has drifted into the view that all natural phenomena ought to be reduced to motions. Light and heat and sound are all due to wave-motions, which travel from the body emitting them to the person who sees light or feels heat or hears sound. That which has the wave-motion is either ether or 'gross matter', but in either case it is what the philosopher would call matter. The only properties which science assigns to it are position in space, and the power of motion according to the laws of motion. Science does not deny that it *may* have other properties; but if so, such other properties are not useful to the man of science, and in no way assist him in explaining the phenomena" (13). This is a perceptive comment, expressing some natural perplexity about the exclusive concern with motion in physics—though I am not sure the scientists have simply "drifted" into it semiconsciously (see chapter 4 for more on this).

[21] I actually question even this apparent possibility in the next chapter, but the general point remains: why *must* all physical forces revolve around motion and change of motion, when there are apparently so many other properties of physical things?

(ultimately) of particles to affect the trajectories of other particles. Mass, arguably the central notion of physics, is typically defined in terms of motion, since it is explained by the notion of inertia, i.e., how much force is needed to accelerate the massive object. In String Theory, the basic explanatory concept—the vibration of strings—simply is a concept of motion, so that variations in motion of the ultimate constituents of the world lie at the root of all physical distinctions. Physics might as well be called the science of motion, and indeed has been so called. It is concerned with matter, *in so far as it moves*.

Yet there are other physical properties of objects—such as size and shape—that are not defined in terms of motion. Motion is a subset of physical properties—just one of the primary qualities—and extrinsic to matter as such. I am struck by the fact that treatises on particle physics never say what *shape* the particles have, and whether different kinds of particles might have different shapes. In diagrams they are usually depicted as spherical, but such a determination never plays a role in the theories of particles—unlike questions of charge and mass. Would it matter if an electron had a star shape? The gravitational force of an object is proportional to its mass alone: that a planet is round, say, plays no role in the theory of its movement under gravity (though it might result from gravity). Physics is supposed to be the science of the properties of matter, but shape seems ignored by the basic laws of physics, which all concern motion. Differently shaped things can all share the same trajectory, and trajectory is all that matters.

This is clearly connected with a common refrain about physics: that it doesn't really say what matter *is*, only how it behaves.[22] The laws tell us what matter will *do* in the way of motion, but they are silent about what it is that moves. What *is* matter? It consists of particles with various masses and charges and spins, but what *are* those particles—what is it to be material? In practice, the physicist will understand matter by contrast to space: it is what exists *in* space without *being* space; or it is what *travels* through space. But if we insist on asking what it is intrinsically, we will be told either that that is an illegitimate question or that we simply don't know (and don't much care). What we have in physics is a kind of motion-oriented functionalism about matter (a type of "behaviorism"): matter is what *acts* in such and such ways in response to so and so forces, where the actions are motions; what it consists of intrinsically is left unspecified.

This leaves it open for the metaphysically minded to step in with suggestions to fill the descriptive gap: maybe what matter is intrinsically is *mental*—as with the doctrine of panpsychism. The intrinsic nature of matter is "experience," while its functional properties are captured by the laws of physics proper. That this type of view gets purchase at all shows that physics is silent on what matter *is*, while telling us an awful lot about what it *does* in the way of motion. We are told that the basic elements of nature are vibrating strings, say; but when we ask what these strings are in them-

---

[22] See chapter 2 and passim.

selves—what they are made of—we don't get a real answer: they are *what vibrates*. But what is matter *really*? we ask, even once we have heard the physicist tell us his wonderful story about the feats it accomplishes. The reason physics can safely ignore such questions is precisely that it is concerned exclusively with motion—and we can talk about and measure motion without worrying unduly about the intrinsic nature of that which moves (at least, so it is assumed). The focus on motion gives us a selective and partial view of the nature of matter, in which its other properties fade into the background or are systematically ignored.[23]

A hardheaded physicist might respond that such questions as the intrinsic nature of matter belong to metaphysics, not to an experimental science. Physics, she might explain, deals with operational definitions, and these precisely involve what matter does, not what it is intrinsically. The reason physics is so obsessed with motion is simply that motion is what matter does—and operationalist physics must be about what matter does. Hence, electric charge is defined in terms of attraction and repulsion, and mass is defined in terms of inertial force. I think my imagined physicist is basically right: to be an experimental science, physics must be operationalist, and operationalism will lead to a focus on motion—but I don't think this shows that the question of the intrinsic nature of matter is meaningless or empty. What it shows is that physics is not a complete science (or study) of matter: to complete it we need to do metaphysics, or whatever we call the subject that takes up where empirical physics leaves off. We must supplement operational definitions with intrinsic or constitutive ones. Or if we can't succeed in that effort, we should concede that we don't really know the whole truth about the nature of matter—we know what it does in the way of movement but we don't know what is. What I am against is the assumption that nothing about matter can be real that doesn't show up in theories of its movement. I am against what might be called "motion functionalism" about matter (or charge, etc.).

Let me put the issue as starkly as possible: Does the world essentially consist of motions and motions alone or is the concentration on motion in physics an artifact of its operationalism, which arguably is required for its status as an empirical science? Have we discovered that the essence of the material universe is just motion or is it that physics deals selectively with motion because of self-imposed (but possibly justified) methodological restrictions? I incline to the latter view, simply because there has to be *something* that moves—there can't be "bearerless motion." And the

---

[23] If it is maintained that we know what motion *is*—we have clear and distinct ideas of it—so that it passes the test for intelligibility and clarity, then the reply will be that this is not really so. We don't have an adequate conception of the essence of motion, only a structural description, some shaky philosophical theories, and an awareness of its effects on our mind. So we can't justify the focus on motion by claiming that motion is an oasis of transparency in a murky sea. Of course, this is compatible with insisting that we have clear and distinct ideas of our *sensations* of motion—we know their nature perfectly well. Then it might be supposed that we should replace motion with sensations of motion as our prime *explanandum* in physics. But that way idealism lies.

question "What is matter in itself?" is not meaningless (which is not to say it is answerable). All we really know is that matter isn't space, as a matter of deep principle, but we have no positive account of what it is.[24] Motion is obviously extremely fundamental to the universe we inhabit, but its salience in physics results not merely from that truth but also from methodological requirements. The very methodology of physics skews its picture of the nature of the physical universe.

I think, then, that there is a kind of double narrowing going on in physics, centering on motion, as a consequence of essentially epistemological scruples, which results in a distorted picture of the nature of the physical universe. First, there is the restriction to the notion of relative motion, which stems from verificationist assumptions: nonrelative motion is also a reality, undetectable and incomprehensible as it may be. There is simply more to motion, as it exists objectively, than change of relative position. Second, the focus on motion itself results from the epistemological commitments of operationalism: if operational definitions serve to make the phenomena thus defined measurable and observable, they also deflect attention from the intrinsic nature of what is so defined. But that nature still exists as an ingredient of reality, hard to fathom as it may be. There is more to matter and motion than these epistemological and methodological restrictions permit us to appreciate. In effect, physics gives us the kind of biased view of matter that behaviorism gave us of mind. After all, twentieth-century behaviorism in psychology was partly stimulated by Bridgman's operationalism in physics; and the latter is just as faulty as the former as a guide to what the world is really like. Philosophy can help in diagnosing how the methods of physics might bias the picture of the physical world that it presents.[25]

---

[24] Even that we have to be careful about: as I suggested in chapter 1, it is arguable that matter is a rip in space, and so a form that space can assume. And some physical theories find no firm distinction between matter and space at an ultimate level. Still, if anything is clear, it is that dense matter is not the same thing as empty space: matter is certainly not a *vacuum*. Even if matter dissolves into fields, and fields pervade all of space, it doesn't follow that matter cannot be distinguished from empty space. Matter is what occupies space; space is what is occupied by matter: that at least seems firm.

[25] There is a general point here about how a discipline's chosen methods can bias its view of its subject matter. The method fixes the practitioner's view of what is real or serious in their domain of enquiry. That syndrome is very familiar in some of the shakier social sciences, e.g., economics, but it is equally true of physics: the form and content of physics reflects the methods adopted as much as the objective nature of the subject matter. Psychology once ignored its own proper subject matter because of (misguided) methodological restrictions; physics, too, might be said to turn a (partially) blind eye to its actual subject matter, viz. the mind-independent, method-independent, measurement-independent world of material objects in space, subject to objective forces and conforming to objective laws. Physics sees physical reality through the lens of its chosen methods, and these inevitably reflect a human perspective—not merely objective reality as it exists in its width and fullness *sub specie aeternitatis*. That there is such a world I don't doubt, nor do I doubt that physics offers an accurate (but partial) view of it; my point is that its methods may not be capable of delivering the total picture of how that transcendent world is in itself—of its real fundamental objective nature.

4 Motion, Change, and Physics

In his article on motion in the *Encyclopedia of Philosophy*, the distinguished historian and philosopher of physics Max Jammer writes: "The predominant role of the concept of motion in physical science poses a problem of great importance to philosophy. Why is it that all processes, laws, and formulas of physics—and modern physics is no exception—ultimately refer to motion, and why is it that even problems in statics, the science of equilibrium and absence of motion, are solved in terms of fictitious motions and virtual velocities?" (399).[1] I think Jammer is onto a good question here, and I propose to answer it in what follows.

Evidently, Jammer regards the ubiquity of motion in physics as both true and surprising. I agree on both points. It is surprising, first, because on the face of it the physical world contains many more properties than those of motion; it appears much richer than that. There is color, sound, shape, size—all the primary and secondary qualities. The physical world seems to consist of far more than amorphous masses in motion—and yet physics is obsessed with motion alone. Physics concerns itself narrowly with the laws of motion, omitting all the rest. Second, physics, in its theoretical core, deals with several kinds of force—with accompanying variation in laws, causes, and explanations. There is not just gravitational force, which deals with projectiles moving through space under the influence of mass; we also have the electromagnetic force that deals with waves propagated through space—as well as the

---

[1] The article is in Paul Edwards, *Encyclopedia of Philosophy* (MacMillan: London, 1967), vol. 5–6. Compare this passage with the like passage quoted from Russell in chapter 3, note 20.

so-called strong and weak forces. The quantum world and the classical world are subject to different principles (with a real problem of unification), which are quite different in essential nature—radiation in one, gravity in the other. Yet all this variety comes out in the wash as motion: all the forces express themselves as motion, and motion is what is common to the quantum and classical worlds. Offhand, one might have thought that different kinds of force would have different kinds of upshot; and that the world of the very small should differ in some profound way from that of the very large: but it turns out that it is all just one kind of motion or another. The forces of nature all work uniformly to propel bodies through space over time. Physics is the science of *displacement*—change of spatial location, going from one place to another. And that's it. Every physical quantity manifests itself in the form of motion, according to physics, despite the underlying variety of quantities. Physics is monistic, regarding everything as motion-related. That is surprising.

But it is also true. Masses move through space, falling and orbiting, under the influence of gravity; electromagnetic waves ripple through space from one place to another; particles orbit other particles; particles spin; particles are released and absorbed; photons travel at maximum speed; quanta jump; strings vibrate. Electric charge comes down to motions caused by attractive and repulsive forces. Heat, as every philosopher knows, is molecular motion—literally the movement of molecules in space. Day and night result from planetary motion. The tides are caused by the motion of the moon in relation to the earth. The material universe is expanding through space. Galaxies are flying apart. Black holes suck matter into them. Even the most static-looking objects conceal a microscopic world of frantic wriggling activity. Physics and astronomy are forever discovering that things move that you thought didn't. It is hard to think of anything in physics that is not ultimately about motion—its modes and rates.

Two attitudes towards this theoretical ubiquity are possible. The first attitude is that it is merely an historical or methodological artifact: it results from the history of scientific thinking, which began with kinematics—the movement of geometrically defined objects through space. Or it stems from the methodological operationalism endemic to physics—defining things by means of their measurable effects, motion being the most salient. The other attitude is that we have here a reflection of the deep structure of the world—that motion is its intrinsic essence. Ontologically, motion constitutes the ultimate Nature of Things, so that physics has simply discovered and recorded this basic reality. To be is to move (or to be capable of moving). Is motion, then, just the selective lens through which we contingently look at the world in doing physics, or is it what the world is all about objectively? Is the motion-centric character of physics a sign of its blinkeredness or its metaphysical verisimilitude? As Jammer rightly observes, this is a question of great philosophical importance.

I propose to argue for an intermediate position, though my view is closer in spirit to the second of the two alternatives. I don't think that all *facts* are (reducible to) facts of motion, even all the physical facts; but I do think that all *change* ultimately comes down to episodes of motion. I certainly don't believe that matter itself reduces to motion: matter is the thing that moves, not the movement itself. Nor do I wish to advocate a reductionist position with respect to physical properties (I will come back to the mind and motion later), construing them as all identical to properties of motion. But all this is consistent with supposing that all change comes down to events of motion—that the mechanism of change is always motion. I could put my position by saying that all change is *constituted* by motion or is *supervenient* on motion—at any rate motion is always at the root of it. That is, whenever an object changes it does so *in virtue of* the motion of itself or its proper parts. All change is ultimately a change of position (of the parts). We might call this the Displacement Theory of change.[2]

The thesis sounds radical, like one of those ancient Greek pre-Socratic flights of fancy ("All is motion"), but it is actually fairly easily established. Take shape and size: if an object changes its shape or size, its parts change locations—it comes to occupy a different volume of space from its initial state. If you squash a sphere into an ellipsoid, you make the object shift its spatial position, moving the top and bottom closer together and the two sides further apart. The particles that compose the object accordingly travel through space to take up different locations. Increasing an object's size, say by inflating it with air, obviously involves making the parts of the object move outwards in space.

Or consider biological growth, of plants or animals: it is easy to see that particles or energy from outside move into the interior of the organism, are processed there, and result in a new form and volume—an expansion through space. The circulation of the blood, say, is simply a type of motion that relocates bits of matter, sustained by the movement of the heart as it pumps, itself a type of motion. Objects are made of particles, and those particles move constantly, producing changes of form and substance. The kinds of change wrought by earthquakes, explosions, ocean waves, and wind are all just changes of location on the part of constituent elements: the motions of one set of particles change and as a consequence other particles are disturbed—generally by simple collision. Smashing an object is simply sending its parts away from each other in space. The extinction of the dinosaurs, itself a matter of the redistribution of molecules through space, is said to have been caused by the motion and impact of a large meteor. The whole evolution of species, like the development

---

[2] Change might roughly be defined as acquiring new properties over time (with some necessary restrictions on what kind of properties count: an object doesn't change if it goes from existing in a world in which I am thinking about lunch to a world in which I am thinking about dinner—say, an object in a galaxy distant from me). It is by no means obvious that all genuine changes result from changes of location on the part of suitable objects, which is what motion is. This is a substantive metaphysical thesis, not a trivial linguistic analysis of the word "change." The *OED* defines "change" as "make or become different," which is pleasantly succinct; but nothing in this definition says that all kinds of becoming different are reducible to alterations of location.

of the physical universe, is a matter of particles moving into new locations: nothing can happen, evolutionarily, without particles taking some sort of trip (natural selection is a kind of particle allocation mechanism). The atoms in your body have no doubt taken a long and winding road down through the eons, moving from one place to another, part of an ancient rabbit here, a prehistoric toad there—until they end up in (say) your left ear lobe. Of such epic journeys the history of the universe is made. History is particles moving hither and thither (or is supervenient thereon). We are all here because atoms moved in certain directions at some time in the past. (This is not to say that the laws of biology are reducible to the laws of motion; I am speaking merely of the underlying physical mechanism of biological change.)

It might be objected that changes of color don't come down to changes of location: when an apple goes from green to red it doesn't need to move and neither do its parts—it just sits on the tree. But such a change would involve a change of reflectance properties in the object's surface, and this is precisely a matter of which parts of the spectrum are transmitted back through space to the observer's eye—as well as involving an intrinsic alteration in the fine texture of the object's surface (microscopic shifts in chemical pigments and the like). If it is protested that an object could change in its color *appearance* without such intrinsic physical changes, because of changes in the way an observer's eyes register the same wavelengths, then the obvious response is that the brain and retina of the observer must have changed—and this too is a matter of the motion of the minute parts of those organs. The action of the brain, in particular, consists of the movement of ions across synapses—and that is clearly a type of motion. Changes of mental state always involve such microscopic motions.[3]

Chemical interactions involve the bonding and separation of elements, which is just the spatial rearrangement of atomic parts; and such processes as freezing and melting are a matter of constituent elements suffering changes of motion—decreasing or increasing, respectively. It is hard to think of any physical process that does not rest upon such changes of location at some level. Energy itself moves from one place to another, driving the universe along. All physical change is either movement *of* an object or movement *in* it. Macroscopic change can result either from

---

[3] The standard thesis of psycho-physical supervenience might then be rephrased as follows: no mental change without a change in the motions of the constituents of the brain; and no two subjects can differ in their mental states or processes without differing in the motions occurring in their brains. If two brains are exactly alike in their constituent particles and in how those particles are moving, then they must be exactly alike in their state of consciousness. If I go from having one thought to having another, this must be because of motions of matter in my brain—such as the flow of ions across synapses. Changes of blood flow to different parts of the brain are clearly just the motions of blood through the veins and arteries. This motion-specific supervenience seems like a far more substantive and controversial claim than the usual general talk of "brain states" as the supervenience base: it seems quite surprising that states of mind should be so dependent on the motion of particles (remember that electrical impulses are the motion of electrons along a conductor—which is what nerve fibers are). Still, the thesis seems true.

macroscopic causes or microscopic ones. In the macroscopic case, the role of motion is generally evident—as, say, when one object comes along and bumps into another. In the microscopic case, the role of motion is not evident to the naked eye, so we are inclined to overlook it. The motions of germs, say, are not obviously the cause of perceptible illness, as when they flood the blood stream like an invading army. However, microscopic motion is just as powerful and prevalent as macroscopic, despite its invisibility, and is the source of much of the change we normally observe. Perhaps if we could perceive microscopic motion as clearly as macroscopic motion we would not need any persuading that all change is engineered by motion.

At any rate, the distinction lacks metaphysical substance, being merely a reflection of our contingent senses and current technology. The motion of the universe is not divided *objectively* into the microscopic and macroscopic.[4] The distances covered by microscopic entities may be small (though this is an anthropocentric description) but the amount of motion can still be considerable, since there are so many of them moving. The important point is that there are not two ontological *kinds* of motion—microscopic and macroscopic—there is just motion *tout court*, and it is responsible for all the change we observe (and don't observe). The world perhaps gives us the *illusion* that not all change is motion-based—as, for example, with the change of color of a chameleon—but to more penetrating eyes the role of motion becomes evident. The change from night to day also seems motion-independent to the naive mind—it just seems to get lighter in the morning—but on closer inspection we learn that this change results from the motion of photons leaving the sun and reaching the earth. Likewise for aging, rusting, blushing, ripening, heating, hardening, dying, thunder clapping, and so on. We have discovered that all such changes are accompanied by shifts of position somewhere by something. In the lingo: all change is supervenient on change of location. Since motion is change of position, we have the result that all change reduces to a single type of change—the thesis of Change Monism.

If all change is rooted in motion, there will be consequences for the *rate* of change. There will be constraints arising from the time it takes to move from one place to another and the time it takes to initiate movement. Thus in the case of heat, say, the rate of change of heat will depend upon how quickly molecules can have their motion altered, as well as on how quickly they can travel (the faster, the hotter). Change will not be instantaneous, since inertia must be overcome and distances must be traversed. Nothing can travel faster than light, let alone at infi-

---

[4] The whole distinction between the microscopic and the macroscopic is clearly anthropocentric, turning upon what we can perceive with our senses. Nothing is *inherently* microscopic. Atoms are macroscopic relative to their elementary constituents, protons, electrons and neutrons—while planets are microscopic relative to galaxies. The entire universe of matter is microscopic relative to all of infinite space (if there is such a thing). The microscopic and the macroscopic are not natural kinds. Thus so-called microscopic motions are not different in kind from macroscopic motions, *just* by virtue of size (they may differ because of the operative forces).

nite speed, so change must be constrained by this limitation on the motion of masses. We would predict that change will tend to be gradual, not abrupt, since motion itself takes time—displacement in space is a temporal process. The notion of change itself is not thus temporal—we can conceive of instantaneous change—but the underlying process of change of location is inherently temporal. The growth of an organism is slow (by human standards), but then the motions that have to be completed for it to happen are considerable. Even the illumination of the earth by the sun takes up a decent chunk of time, as light makes its (roughly) eight-minute journey from sun to earth. The octopus may exhibit a sudden change from one color and texture to another, at least to the naked eye, but this happens only because of the finite velocities of the pigments and molecules that effect such change. If particles could shift from one place to another with infinite speed, change could be instantaneous: but it is a deep fact about the world that such a thing is impossible. We must always wait for matter to make its leisurely way across space for anything new to happen. Motion is necessarily time-consuming, so change necessarily takes a while.

The role of space in change also stands forth: there cannot be change in the absence of space, because motion simply is change of spatial position.[5] Change is always the exploitation of the resources of space (particularly what I earlier called its penetrability). This is obvious for the kind of change we call motion, but it turns out to be essential to change more generally. For change to occur, space must be occupied successively at different locations by discrete entities. Objects (or their parts) must have a varying relation to space in order that they should change in any way. Thus time and space are brought into the very core of change, because of its dependence on motion (which is defined in terms of space and time). Change occurs because of the receptivity of space to fresh occupation over time.

Is there anything that doesn't move? If so, we would expect that it doesn't change. Numbers don't move: the number 2 cannot travel from one region of space to another. Nor does it have parts that can displace through space. Numbers are not located in space, so are not even candidates for movement. And numbers are changeless. If numbers could change but not move, then we would have a counterexample—but they do neither. Space is similar: it cannot move, but it also doesn't change—not really. Parts of space cannot traverse space, and they are magnificently unchanging. It is true that when a new place is occupied by an object, space enters into a new relation with matter, but no intrinsic change is thereby produced: the "change" is purely

---

[5] We can see here the outlines of an argument against Cartesian dualism: minds change, but no movement of material parts in space can be responsible; and immaterial substance does not have parts that can move in space, since it is without extension. How does Descartes explain changes of mind, when mind is so cut off from motion, the only source of change?

relational.[6] There can be no change *in* unmoving space, and space itself is incapable of movement. If, *per impossibile*, its parts could move, then it would be subject to change, but without this it is changeless and (presumably) eternal. Neither do fields of force change space itself: a field may begin to pervade space at a certain time and then cease to do so later, but the underlying space is not intrinsically altered. Space, unlike matter, does not have moveable parts—and that is what makes something a subject of genuine change (as opposed to merely coming to stand in new relations to things: so called "Cambridge change"). Material objects are the paradigms of change, and they also are what move through space: these are not accidentally connected characteristics.

What about mental change? Does this also consist of masses in motion? It certainly seems hard to see how, at least at first sight. Suppose I go from thinking about philosophy to thinking about surfing—is that a matter of a redirection of motion in my constituent parts? This obviously raises the mind-body problem and the question of dualism. Without discussing that problem in any detail, we can say that, whatever a mental change intrinsically *is*, it has a *correlate* in the brain that involves motion: and if we regard that correlate as the mechanism of mental change, then we can conclude that motion also lies at the basis of mental change. (Descartes could in principle have been a mechanist about mental change, though a staunch dualist about mental ontology.) That is, we can restrict ourselves to a supervenience thesis, holding that all mental change *depends* upon motions in the brain—ions crossing synapses and the like. Nothing changes mentally without such motions in the brain, and the same motions produce the same mental changes. Electrical potentials initiate motions of particles in the brain, and without these the mind would be static. Try making mental headway with your brain in motion freeze-frame! There is no thinking without the jiggling of brain particles—without something cerebral moving from A to B. Hence the neuroscience talk of the *transmission* of impulses from one brain part to another. Of course, it is deeply puzzling *why* this should be so—for what have the motions of particles got to do with the flow of consciousness? But, *de facto*, this is how the brain basically works, like any other physical object—particles

---

[6] The astute reader of chapter 1 will sense an inconsistency between what I say here and the theory defended there. Here I say that the motion of matter through space does not change space intrinsically; there I contended that motion *destroys* successive regions of space—those that are occupied by the moving object. But two points: (a) to destroy a portion of space is not to make a change *in* that portion (I don't change you by totally destroying you); and (b) the portion of space is immediately re-created as it was once the moving object leaves it, emerging quite unaltered. Another potential counterexample is provided by General Relativity: doesn't the introduction of a massive object into a region of space change its curvature, and isn't this an intrinsic change in the space? This is hard to evaluate, but taken literally the *bending* of space is a kind of movement (contrary to Newton's claim that no part of space can ever move). If mass causes lines in space to go from straight to bent, this is a kind of motion in the lines of space. But I confess to not being very clear what the curving of space is supposed to consist in. In any ordinary sense, at least, space is not changed by what occurs in it: events and objects leave no trace on the space that once housed them—the history of a region of space cannot be inferred from its present nature.

in motion.⁷ After all, growth of the brain, itself essentially involving motion, leads to development of the mind, so we know that this kind of mental change rests upon change at the level of moving matter. Mysterious as it may be, changes of mind result from changes of position in underlying matter—like all other changes.

This point helps to formulate Descartes' problem of interaction. Why is it so hard for the unextended mind and the extended body to interact? Because changes in the body consist in onsets of motion, and how can these affect (or be affected by) something that engages in no such thing? How could that which is not spatial and has no motion be in causal interaction with something whose essence is matter in motion? Changes in the body must be changes of location of its parts, but that seems the wrong *kind* of thing to affect something that has no spatial parts and is incapable of motion as a matter of principle. Motion can only lead to more motion—but, for the Cartesian, the mind is an unmoving substance. Certainly, the moving parts of the brain cannot *collide* with thoughts in the mind, as Descartes conceives the mind. Nor can we suppose that the action of one on the other can be anything like gravity, since thoughts have neither location nor mass. The interaction, on a dualist picture like Descartes', seems quite unintelligible, because motions in the brain have no way to affect processes in the nonspatial motionless mind.⁸

If what I have been saying so far is right, then the reason that physics is so obsessed with motion is that it is primarily a theory of physical change. Since all change is motion-based, a satisfactory theory of change can afford to deal only in motion. This is rather different from saying that physics is a theory of the nature of matter. As has been argued elsewhere, physics does not really tell us what matter is, so it can hardly claim to be a successful theory of the *nature* of matter. But if it is a theory of *change* in matter, it need only concern itself with the nature of material change, not the nature of that which changes; and change of location is the fundamental change that matter undergoes. The operationalism of physics also starts to make sense, since change is what *happens*—what something *does*—and so the emphasis on operation

---

⁷ We could state the mind-body problem thus: how is it that the motion of particles in *brains* gives rise to consciousness but the motion of particles elsewhere does not? How do electrons traveling down nerve fibers produce consciousness, while the movements of electrons down other types of conductor do not? Aren't the motions the same in the two cases? Shouldn't the same causes have the same effects? What makes the motion of electrons in the brain so different? Particles flow and jiggle everywhere, not just in the brain, so what makes the flowing and jiggling in the brain so unique? (The panpsychist will see here a marvelous vindication of her views: *all* particle movement is accompanied by consciousness, in brains and out. And one has to agree that the situation is anomalous otherwise.)

⁸ Descartes may save the mind from mechanism, but the price is that he has no theory of mental change—since interactions between the mind and the brain are impossible and the mental substance itself has no moveable parts. He must presumably appeal to brute unexplained mental change. This is a bit like the problem of how God can change, given his nonspatial composition (he has no moveable parts)—except that God is not *supposed* to change. But minds change all the time—and Descartes has no account of how this happens. Certainly, it cannot be by the same means that bodies change. He *might* try claiming that minds don't really change—it is only an illusion that they do—but that seems a hard road to hew.

in physics is perfectly intelligible ("Don't ask what an electron *is*, ask what it *does!*"). In physics, the beginning (and end) is the *deed*—and the deed consists in traveling from one place to another. Motion is how matter does things, how it brings about change, so it is natural to study motion if material change is your central interest. But if you are interested in matter as such—what it intrinsically is—motion starts to seem less than crucial (can't there be static matter?). Motion may be extraneous and contingent to matter itself (not part of the essence), but it is hardly extraneous and contingent to the *activities* of matter. It seems to me, then, that physics is exclusively concerned with motion because it has fashioned itself as a theory of physical change—and all such change really comes down to motion.[9] Physics is really about how the universe goes from one state to the next—which is not the same as the question of what kind of thing the universe is in itself. Motion is how the universe *evolves* over time, so a science of that evolution will be a science of motion. Matter, then, will be seen simply as the *agent* of motion—whatever that may be.

The Displacement Theory of change has an interesting corollary, namely that the universe is fundamentally unchanging. Change of position is a very minimal kind of change: when an object—particle or person—goes from one place to another nothing changes intrinsically in that object simply as a result of the change of position. The change is entirely relational or extrinsic. This is true both for the positions involved and for the objects that take them up: neither is changed in its nature—only in the relations between them. Changes of position are "Cambridge changes." It follows that if all changes are changes of position, then all changes are ultimately merely relational. All intrinsic changes are relational changes with respect to the *parts* of the object in question. Change of shape, say, is intrinsic to the object, but it consists in changes that are purely relational, since the particles that move do not themselves change shape. In a sense, change of shape is a superficial phenomenon, resting upon constancy of shape in the constituents combined with variability in position. The ultimate parts of things only ever change their position, but that is enough to produce intrinsic changes in what they compose.

So to say that motion is the root of all change is to say that no change is ever fundamentally or irreducibly intrinsic: all intrinsic change cashes out as relational change. The universe is thus obeying a principle of parsimony in how it contrives

---

[9] You might reply that physics is also about the architecture of matter, its structure, breaking bodies down into their constituents—from molecules to atoms to elementary particles. This is not about change and activity per se. True enough: but the entities that compose these configurations are themselves conceived in terms of motion—that is, by how they interact with other entities in space. The electron, say, is understood as that which orbits the nucleus in a certain way and attracts and repels other particles according to certain laws—-as well as having various absorption and diffraction properties. It is defined by its "motion profile." Its electric charge is understood precisely in terms of its effects on the motions of other particles. The architecture of matter is therefore conceived as a collection of moving parts, represented as such: it is a structure of mobility, of propensities to movement. The architecture postulated is dictated exclusively by observations about motions. There is nothing static about the atom.

change—it gets the most from the least. It is conservative with respect to change, changing as little as it can. At bottom everything stays the same, even as things rush around from place to place. There is much energy in the moving, but there is laziness about effecting internal change. All the novelties we observe in the universe, of configuration, color, texture, and so on, are the result of new paths taken by unchanging particles. All intrinsic changes are supervenient on relational changes, specifically changing relations to space (or other objects, if you only believe in relative motion). The empirical world is more Platonic than Plato thought—less of a perpetual flux. Ultimately, things stay the same, while engaging in an enormous amount of movement. It is permanence that gives rise to impermanence, courtesy of movement.

It is a question whether the dependence of change on motion is a contingent fact we have discovered about our universe in the course of empirical inquiry or whether it has some sort of a priori and necessary status. Are there worlds in which the Displacement Theory is false, where intrinsic change goes all the way down and motion is not the universal source of change? Such a world is easy to describe without contradiction, but is it really possible? How metaphysically deep does the dependence of change on motion go? Is it a basic structure of reality as such? I am inclined to think it does go deep, but mounting an argument for this is by no means trivial.

First, it would be surprising if in our world the thesis held universally and systematically and yet had a completely contingent status: it can hardly be an *accident*. Despite the variety of entities, forces and laws in our world, all change turns out to involve motion—that suggests something inescapable about it, as if the universe can't help behaving this way. Second, it seems reasonable to suspect that the conservative nature of change on this view is of metaphysical significance: motion is the basis of change *because* it exemplifies the principle of purely relational change—no intrinsic change at the most basic level. Suppose intrinsic change at the basic level was impossible, perhaps because it would be inexplicable; then the way to get intrinsic change at the higher levels would be to exploit the nature of motion, which allows for macroscopic change without intrinsic change in the basic elements. All intrinsic change is derivative from nonintrinsic change at the basic level, so we don't have to face the fact of genuine change at the foundation of the universe—and change as motion allows this. Thus we can explain change without having to take change as primitive, by appealing to motion; radical change does not have to be possible for intrinsic change at the level of composite objects to exist. Motion permits this to be so. It is not clear that anything else could secure the conservativeness of change. I don't have any demonstrative argument for why intrinsic change cannot be fundamental, but I think it is an intuition with some force, and making change depend on motion is a way—perhaps the only way—to avoid postulating it.[10]

---

[10] Change of shape or size necessarily requires motion of parts, because of changes in relation to spatial occupancy—the matter has to go from here to there. But change of color doesn't seem inherently motion-dependent: couldn't it occur without any motion of particles, but just primitively? Isn't

If there is an a priori component to the thesis that change is always mediated by motion—perhaps because of the conservativeness of that mechanism—this throws new light on the status of such familiar philosophical examples as "heat is molecular motion" and "light is a stream of photons." These are proposed as instances of the necessary a posteriori, with the latter characteristic taken as uncontroversial. I don't dispute that such propositions are empirically discovered, but our present considerations suggest a subtler picture of their epistemology. Consider the characteristic changes associated with heat and light (or with water—freezing, boiling, etc): increases and decreases of temperature, the falling of light onto a dark surface, say. If all change is rooted in the motion of parts, and this is an a priori truth, then we know a priori that these kinds of changes involving heat and light *must* involve (hidden) motion of *some* sort: heat must be made up of something that moves in such a way as to increase temperature; and light must be made up of something that likewise moves when changes of illumination occur. Granted, we don't know a priori what specific kind of motion might be in play, and of what entities; but we do know that the underlying mechanism has to involve motion of one kind or another. The proposition expressed by "heat is motion (of some sort)" *is* a priori.

Similarly with the freezing and boiling of water: water has to have parts, presumably hidden, that move in certain ways when these changes occur—though the precise nature of these moving parts awaits empirical determination. This is nontrivial knowledge, and it follows from a philosophical thesis that appears to have an a priori rationale, even if not an apodictic one. Or perhaps we should drop the traditional epistemological dichotomy of the a priori and a posteriori and say instead that the involvement of motion in these kinds of change issues from a principle of the highest generality. It is not that we come to the question of what heat is without any prior knowledge about what it may be: we know that change involves motion quite generally, so that in this particular case the only empirical question is what precise type of motion on the part of what kinds of constituents we have. The role of motion in enabling change is not gleaned solely from individual cases such as these by induction, but rather issues from far more general considerations concerning the very possibility of change—at least if I am right about the obligatory conservativeness of change *au fond*.[11]

there a possible world where objects spontaneously change from red to blue and nothing in that world moves? That doesn't sound logically impossible, and yet it has about it the whiff of the miraculous—there is just no *account* of the change of color. But when objects change because of underlying motions nothing miraculous needs to be going on—presumably because then there is no intrinsic change at the most basic level. Change of position is not intrinsic change, and if it can explain intrinsic change then nothing miraculous needs to be postulated. But a brute change of color, not mediated by any changes of motion in anything, seems magical—in contrast to the changes of color we observe in the actual world, which do involve motions and reconfigurations of minute parts.

[11] Experience and reason work more closely together in these kinds of cases than the usual dichotomy of the a priori and a posteriori allows for. Some knowledge has a strong rational component, while also having an empirical component—as with very general features of the physical world

Then too there is the further question of whether the Displacement Theory logically implies atomism. I have certainly formulated it in atomistic terms, this being such an entrenched part of our world view (I mean atomism in the broad sense that includes field theory). If all change depends on motion—hidden motion if necessary—then we need entities that move, and these will often be parts of the object that changes. For a plant to grow, say, the plant must have (or acquire) parts that move, even when they can't be discerned with the naked eye. To this extent the nature of change leads us to an atomistic conception of things, or at least a mereological conception, in advance of any detailed experimental inquiries (is this how Democritus came up with the idea?). However, it is not clear that the parts in question have to be *discrete*, as atoms typically are. Could we not accept the picture of moving parts generating change while holding that matter is continuous in structure, not composed of discrete and spatially separated elements? Thus when an object changes shape its parts move through space, but these are like geometrical parts, continuous with adjacent parts. We squash a sphere into an ellipsoid, thus causing its parts to move in relation to each other: but it doesn't directly follow that these parts are discrete and separate—maybe the matter in question has no interstices. Couldn't continuous matter be elastic? The idea of moving parts does not *entail* the idea of moving discrete parts. So our current atomistic picture (or that of Democritus) doesn't follow from the very nature of change, though a weaker kind of atomism does—the idea of composing (but contiguous) units, geometrically distinct, that travel through space to effect change.

Annihilation at the macroscopic level is the dissolution of objects into their parts and the dispersal of these parts in space. This is a type of motion, corresponding to the change in question. But what of the parts themselves—how are they destroyed? If all change is motion, and annihilation is change, and there are objects without parts, how can such objects cease to exist? A thing cannot cease to exist simply by

(such as conservation principles, the connection between motion and change, the contrast between matter and space). Kant employed the category of the synthetic a priori, but it might be better to use the concept of the "rational-empirical"—where the relative contributions of reason and experience can vary through degrees. Sometimes reason plays a preponderant role, but sometimes experience is preponderant—as with straightforward historical or geographical facts. Our knowledge that heat is the motion of molecules has both components: reason tells us that heat must involve motion of some sort, since all change does; but experience is needed to determine what kind of motion and of what. As I shall suggest in chapter 5, the law of inertia has a strong reason-based component, though it can receive empirical support from experience. Physics seems full of such mixed cases, with reason and experience tightly linked. The usual examples trotted out to illustrate the distinction between a priori and a posteriori knowledge don't fall into this grey area—such as analytic or mathematical truths and truths about singular concrete facts or laws of nature. But there do seem to be many cases in which the old dichotomy is too stark and simple-minded, such as high-level principles of physics. To anticipate the topic of the next chapter, what kind of truth is it that an object at rest will continue at rest unless disturbed by an outside force? Such a proposition is neither like "Bachelors are unmarried males" or "2+2=4" nor like "The Thames flows through London" or "Gravity obeys an inverse square law." Intuitively, it seems both recommended by reason *and* confirmable by experience.

moving; it has to break down into parts that move away from each other, ceasing to form a whole. But the ultimate constituents of matter cannot do that, since they have no parts. It follows, then, that they cannot be destroyed. More exactly, either the ultimate particles are eternal or they can be destroyed in some other way than by motion of their parts, thus contradicting our thesis. I note that in physics the immortality of ultimate parts is generally assumed (often motivated by conservation principles)—a photon, say, never goes out of existence.[12] My point is that this follows from the general considerations I have been adducing. Suppose the ultimate entities are strings: assuming they have no parts, we can say that the only changes they are subject to correspond to their pattern of vibration (a type of motion); they cannot cease to exist, because they have no parts that can move away from each other and be dispersed. So the thesis that all change is motion, including annihilation, implies that there are primitive elements that are eternal—as well as implying that intrinsic change is always superficial. At the foundation, things don't change and they go on forever.

A final point must be made, referring back to our earlier discussion of the nature of motion.[13] I have been claiming that change consists of motion, but I haven't said anything substantive in this paper about what motion is. In the previous chapter I discussed the nature of motion at length, finding inadequacies in both relative motion and absolute motion. I concluded that the best view might be that motion is something we-know-not-what: the word "motion" refers to something real and objective, but we have no good idea of what that might be like. Motion is a natural mystery. Now I can combine this line of thought with the argument of the present paper: change consists of motion but motion is mysterious. Therefore change is mysterious. If we don't really know what motion is—we have no positive conception of its intrinsic (nonabstract) nature—then we don't grasp the nature of change

---

[12] This may be an oversimplification: the mass or energy of a photon cannot be destroyed, though the individual photon may be absorbed by another particle. That is, the *stuff* of the photon cannot be destroyed—or that of any other basic particle. I would regard the packet of energy composing a photon as a type of particular, in the philosopher's sense—so that this particular is immortal. It is a delicate question whether the original particle composed of this stuff is really destroyed, as opposed to changing its form in certain kinds of interaction. In any case, the basic building blocks of matter (let's call them packets of energy) are immortal—and they are particulars, not universals, concrete denizens of space and time. This is one of those remarkable facts, mentioned in the Introduction, that are often just reported as mundane realities. But think about it: there are physical particulars pinging about space at maximum velocity, colliding with things, bouncing around, changing direction, light as can be, and they will be around *for all eternity*. Nothing can end their space-traveling existence. They may be a photon today and part of an electron tomorrow (in the form of higher energy), but they are true eternals. The mighty sun will be gone in the blink of an eye, but that photon (to oversimplify slightly) will be around for ever—whipping from one end of the universe to the other, secure in its being. A human life is less than a nanosecond compared to the life of a photon. They are the permanent fixtures of the universe, like space itself. Photons (packets of energy) will be around as long as God.

[13] See chapter 3.

either, since change reduces to motion. A Newtonian who accepted the Displacement Theory of change would take himself to grasp the essential nature of change, since he takes himself to know what motion intrinsically and intelligibly is; and similarly for a Leibnizian who accepts the Displacement Theory—for him, motion is change of relative position. But for someone who finds both viewpoints unsatisfactory, and suspects that motion belongs with matter on the mystery side of the ledger, the indicated conclusion is that change is not intellectually transparent. Or alternatively, our only handle on the nature of change is provided by the merely structural-cum-mathematical conception of motion supplied by physics.[14] On that conception, we can measure motion, and provide laws for it, but we can't elucidate its intrinsic character. If the present chapter is right, that conclusion carries over to our conception of change: we have the measures and the laws, but we don't really know what change is. We don't, that is, know what change *consists in*. That is quite a big lacuna.[15]

---

[14] We may think we are better off with respect to motion than we are because of the way vision presents motion to us. But that is not a plausible view, as I argued in chapter 3: vision cannot be regarded as revealing the real essence of motion. Still, we are prone to *think* we understand change better than we do because we think that motion is revealed to us in sensory perception. If motion is obscure, however, then so is change: the word "change" denotes a process whose nature we do not grasp—because it denotes motion and we don't grasp motion. We have a grip on change of relative position, and this can feed into our conception of what change in general consists in: but mere change of relative position is not motion, as I argued in chapter 3, so this does not really give us the right notion of change. We have, as Locke would say, an "inadequate idea" of change—or better, of the mechanism of change. Change comes about because of the mysterious process of motion, and hence is itself mysterious.

[15] Let me be clear here: we can *define* change, so we are not ignorant of the meaning of the word—it means, roughly, acquiring new properties over time ("becoming different"); but that does not imply that we know what change consists in *de re*. If the argument of this paper is right, we also know that change consists of motion—and that goes beyond the meaning of the word. What I am saying is that we don't know the *nature* of change, because we don't know the nature of motion—what motion is in itself. (There is also the question of whether we know the nature of time.) We know that change consists in motion *whatever motion may be*—but we can't fill in the blank and produce an account of motion. If absolute motion were intelligible and correct, then we could say what motion is—but in the end that theory foundered. From the fact that we can define the words "change" and "motion," in the dictionary sense of "define," it doesn't follow that we know what the referents of these words *are*. And our linguistic competence should not be allowed to foster a false sense of epistemic security in us regarding the nature of objective reality.

5 The Law of Inertia

Newton's first law of motion states that a body at rest will remain at rest unless acted upon by an outside force and that a body in uniform motion will remain in that state of motion unless acted upon by an outside force. In his words: "Every body perseveres in its state of rest, or of uniform motion in a right line, unless it is compelled to change that state by forces impressed thereon" (19).[1] He glosses the law thus: "Projectiles persevere in their motions, so far as they are not retarded by the resistance of air, or compelled downwards by the force of gravity" (19). The last two forces mentioned—air resistance and gravity—give rise to the illusion that motion does not naturally persevere, but absent such forces motion is preserved (contrary to Aristotle). The second conjunct is traditionally regarded as the crucial one, the first being thought self-evident. More exactly, it says that an object moving uniformly in a straight line will continue moving uniformly in a straight line, without either accelerating or decelerating, to eternity unless acted upon by an outside force. (If an object is initially moving in a curved path under the influence of a force, it will continue in a straight line once that force is removed: it won't preserve the curved trajectory.) Intuitively, an object will keep moving as it is moving unless something stops it. A change of direction or velocity needs the imposition of an external force. This means that in a universe without external forces, but just the moving object itself, motion will continue unimpeded forever (granted the boundlessness of space).

---

[1] *Principia*, "Axioms, or Laws of Motion." Page references appear in the text.

The law also states, as a corollary, that the tendency to preserve motion is independent of mass: no matter how massive an object is (either very light or very heavy) it will continue in its state of motion unless prevented from doing so. Mass determines how much force is needed to change a state of motion (say, from a position of rest to rectilinear movement), but it doesn't affect the ability of an object to keep moving without external influence (a light object will continue just as much as a heavy one and vice versa).[2] Indeed, the law applies equally to the entire mass of the material universe, as well as to the tiniest particles. Bodies resist changes of motion in proportion to their mass, but they do not vary in their propensity to retain motion once initiated. Bodies are not all equally inert, but they are all equally equipped to preserve their state of motion. Motion is in its nature an activity that keeps on giving without any help from outside (or indeed inside).

Newton's conception rejects the Aristotelian view of motion, which is why it was deemed revolutionary (Descartes and Galileo had already grasped the basic point). Aristotle held that motion is an activity that naturally dissipates: while an object at rest will naturally stay at rest without outside interference, a moving object will lose its state of motion with time whether or not anything interferes with it. The view was motivated by ordinary observation—balls rolling along planes, arrows fired into the air. Things spontaneously run out of motive steam, it appears. It took Galileo, Descartes, and Newton to put these effects down to friction and gravity, and to consider the case of movement in force-free empty space. For Aristotle, it took constantly renewed force to keep things moving, and an initial force would dissipate unless supplemented; but for Newton, no force is needed to preserve motion, and only an impressed force could retard it (or speed it up). As we might put it, Aristotle took gradually reducing motion as natural, while Newton took preservation of constant motion as natural. Absence of force underlies uniform motion for Newton; presence of force is necessary for uniform motion for Aristotle. Continued motion is effortless for Newton and effortful for Aristotle. These are clearly profoundly different ideas, reflecting deep principles about causation and change. And, of course, Newton is now thought to have been in the right as against Aristotle.

All this is familiar enough: my question is what the status of the law of inertia is—where does it come from? Aristotle's view obviously comes from ordinary observation—and his description of how things actually move on earth is broadly correct. But he gets it wrong theoretically; he misunderstands the basic principles. How did

---

[2] We tend to think a heavier object will slow more naturally than a light object, because its weight counteracts the initial impetus, but in fact it will persist in its movement in exactly the same way a lighter object does. Terrestrial observation also suggests that very light objects, like feathers, naturally slow down quickly—but in fact this is due simply to the friction created by air. A feather suitably launched into force-free space will keep moving just as a bullet will. Once an object, of whatever mass, is changed from a state of rest to a state of motion by the application of a force, it will continue moving at the same rate—from the heaviest metal to the lightest pollen. The amount of force needed to induce a state of motion is irrelevant to the persistence of that motion.

Newton know that? Not by making more observations than Aristotle and noticing that objects persist in their motion unless prevented from doing so. Rather, his advance was conceptual: he conducted a thought-experiment involving frictionless empty space and zero gravity, and concluded that in such a set-up an object *would* preserve its state of motion. He never actually conducted the experiment, if only because of the inaccessibility on earth of gravity-free space.

The law was then used to explain the observed movements of terrestrial and celestial bodies, by adding in the influence of forces. It seemed evident to Newton, once you reflected on it, that the law of inertia *must* hold—that Aristotle's view *had* to be wrong. There has accordingly been some debate about whether Newton's three laws of motion are empirical generalizations or a priori principles, the feeling being that they are more than merely the (inductively generalized) record of empirical observations. In fact, when Newton enunciates his laws, he does under the heading "Axioms, or Laws of Motion," as if he is stating mathematical certainties. Poincare pointedly asks:

> Is the principle of inertia, which is not an a priori truth [he seems certain of that], therefore an experimental fact? But has anyone ever experimented on bodies withdrawn from the action of every force? And if so, how was it known that these bodies were subjected to no force? Is this because it is too remote from other bodies to experience any appreciable action from them? Yet it is not farther from the earth than if it were thrown freely into the air; and everyone knows that in this case it would experience the influence of gravity due to the attraction of the earth. (66–7)[3]

Perhaps the old dichotomy of the a priori and the empirical is too crude, in the light of propositions like Newton's laws, but I think we can see why it has been felt that there is at least an a priori component to Newton's laws. The first law, in particular, seems to result from an *analysis* of the situation, not from any new empirical findings. Once you understand what Newton is saying, you are inclined to assent to his law: it recommends itself to reason. What I want to investigate is why this is so: why does the law of inertia strike us as self-evident (once it has been properly grasped)? Certainly, the first part of it, about the preservation of rest, seems self-evident (presumably also to Aristotle); but the second too, once it is clearly articulated, has the look of the undeniable. It contrasts markedly with the inverse square law linking gravity, distance, and mass—this strikes us as entirely empirical and lacking in a priori self-evidence. Does the law, perhaps, spring from some more general metaphysical principle that commands immediate assent?

We can start by reformulating the law so as to remove its conjunctive form. What we need is this: the state of motion of a body remains unchanged unless acted upon by an external force. Here we introduce the general notion of a state of motion,

---

[3] *Science and Hypothesis* (General Books, 2009; originally published 1905). He adopts a "conventionalist" view of Newton's laws, which is a kind of positivist substitute for the a priori.

comprehending both rest and uniform motion, and explicitly build in the idea of change. A change of motion includes acceleration and deceleration, as well as a change of direction—anything aside from uniform motion in a straight line. So we can say that changes of motion need an outside cause, i.e., a force to bring them about. Aristotle thought that changes of motion could occur without an outside force, but Newton teaches us that motion cannot change unless something outside makes it change. Change of motion needs an external cause, while constancy of motion does not. Or again, constancy of motion needs no explanation, but change of motion does. Or yet again, constancy of motion is natural to an object, while change of motion is not. Having the same state of motion over time is what objects do if left to themselves. An object has to be interfered with if its state of motion is to be anything other than constant. An object will remain constantly at rest unless forcibly shifted, and an object in motion will retain the same motion so long as nothing thwarts its progress. Constancy of motion is, as it were, the default setting of the universe, prevailing unless overruled by countervailing forces. Constancy needs no sustaining cause, while change requires a disruptive force.

I have repetitiously reformulated the law of inertia so as to make it successively more general and focused on the notions of constancy and change. The obvious next move is to see if we can delete reference to motion in the law and still have a truth. The more general principle is then this: constancy of properties needs no explanatory sustaining cause, but change of properties does. Or again: objects will persist in their nature unless interfered with by outside forces. More pithily: things don't change unless forced to change. The law of inertia is then a special case of this more general principle, and inherits its epistemological and metaphysical status.

But is the general principle true? The question is somewhat delicate. Consider mass: do objects retain their mass unless an outside force interferes with them, say by splintering off a piece? The answer is yes: mass is conserved in an object unless something comes along to change it. Objects don't spontaneously lose their mass any more than they spontaneously lose their motion: it has to be taken from them. Things don't get lighter over time of their own accord any more than they slow down of their own accord. Consider a body moving in free space with no force acting on it (from without and within): will it lose its mass over time? Clearly not, I say. We could therefore state a first law of mass: bodies persevere in their mass unless acted upon by an impressing force. They do not get heavier or lighter unless something *makes* them do so, such as the friction of the surrounding air. This seems obvious, and not in need of experimental verification (though much evidence could be cited in support of the proposition).

Or consider shape and color: don't objects retain these properties over time unless a force interferes with them? Of course, forces do interfere, relentlessly, ubiquitously—just as on earth forces such as friction and gravity do retard the natural movement of objects. An object can be crushed and squeezed and thereby lose its shape, and it can lose its color by being painted or left in the sun to fade. But left to their own devices, in a force-free environment, such as never actually obtains on earth, objects would retain their shape and color: there is no *spontaneous* loss of such attributes—as if they harbor a tendency to expire of their own being and nature. Nothing about being square or red guarantees a loss of these qualities over time without some independent agency to do the work of modification. So motion is by no means alone in its tendency to persist in *its* being and nature. Permanence is the rule, unless externally disrupted. Newton might put his objection to Aristotle as follows: if you don't think mass, shape, and color change spontaneously, why are you so sure motion does? Why should motion be the exception the general rule of property preservation?

Here we must enter a qualification. Someone might object that some changes of mass, shape, and color do come from inside the object; not all such changes are *externally* imposed. Can't there be internal forces in an object that lead it to lose mass or shape or color? Can't the enemy to constancy lie within the gates? Some causes of change surely can issue from within an object's spatial boundaries—as when the particles of an object undergo a convulsion of energy or a shift of position. Can't an octopus change its own shape, color, and trajectory from within?[4] Couldn't a chemical compound explode from inside? Don't planets sometimes get knocked off their course by huge interior upheavals? Wouldn't a human being change dramatically over time as he or she traveled through empty space, with no impressing forces from outside? Of course, these things are all possible, indeed actual.

But notice that much the same is true of the case of uniform motion. Newton's law is generally stated by referring to "outside" forces that interfere with the object's flight path (Newton himself speaks of "impressed forces," thinking mainly of gravity from an outside body): but it is entirely possible for an object's motion to be changed by some event occurring within it—an internal explosion, say. A force could be unleashed from inside it that made it slow down or speed up or change direction. So the reference to "outside" forces has to be a convenient looseness that does not go to the heart of the law. What the law is really saying, of course, is that change of motion requires an *interaction* of some sort—and the interaction could be between an object and one or more of its own internal parts. The idea is that a change of motion requires a *new* force—in addition to whatever started the object in motion: nothing "outside" *this latter* is required once motion has begun. If a new force emerges from within the object, then this could well affect its motion: the force doesn't have to come from some spatially removed object. Aristotle thought the initiating force naturally wears out, so that a change of motion requires no extra supervening force; Newton is saying that the effects of the original force keep on working indefinitely and can only be thwarted by some new force coming on the scene—whether it emanates from outside *or* inside the spatial boundaries of the object. The essential point

---

[4] It could also change its mass, say by biting off one of its tentacles, just as a person can lose mass by going on a self-imposed hunger strike. Clearly, you can change your state of motion by simply deciding to speed up or slow down—and this is not the result of a force impressed from *outside*.

is that change of motion requires an *additional* force to that which got the object moving in the first place—otherwise constancy of motion will eventuate.[5]

But now we can see that an analogous refinement is necessary for the other constancy laws I have mentioned. An object can change its mass, shape, or color because of an endogenous force, to be sure, but any such force is extraneous to *having* that original mass, shape, or color: it is something additional, over and above, "outside." The key point is that having a particular mass, shape, or color is not something that *inherently* dissipates—as if it had an expiry date written into its inner nature. An object can indeed decay from the inside out, which is equivalent to a force operating from within, but this is not a matter of its properties themselves having an inner tendency to weaken or transform or alter in any way—as Aristotle thought the property of uniform motion does. Presumably this is quite obvious for properties in general—they contain no program for their own extinction.

What Newton saw is that the property of uniform motion is no different: it too is endowed with a tendency to persist, to remain constant, not to fade away. This is already clear enough for the property of rest—rest is not itself restless to move—but motion is inherently active (or so we think), so it is more surprising that it too persists in its being unless curtailed from outside. What Newton saw, on this interpretation, is that uniform motion is like mass and shape in not containing the seeds of its own destruction—that it remains constant if not actively subverted. Just as a cosmically isolated object (with no changes in its interior forces) will not change its mass or shape or color, so it will not change its state of motion—even if it is traveling at a million miles an hour, weighs as much as ten suns, and has been in that state of motion for the last five billion years. And if we put the point in terms of particles, we can dispense with the qualification about endogenous forces: an elementary particle will not change its mass or shape or state of uniform motion unless some (spatially) outside force interferes with it.[6]

If that is the right way to look at the matter, then the law of inertia can be deduced from a more general law of perseverance, which we may call "the law of uncaused

---

[5] If no force initiated the movement—the object was just born moving—then the law states that the object's movement only changes with the imposition of a force. However the uniform motion originated, whether by force or not, a change of motion requires something new in the way of force: any deviation from uniform motion needs a force to act at the time of the deviation—which Aristotle did not think.

[6] This suggests a new formulation of the general law: *matter* does not change its properties unless acted upon to do so. The substance of material things, whatever that ultimately may be, is such that it naturally preserves any properties that are currently instantiated. Matter is thus not by its nature volatile or itching to change: it tends towards permanence, stability, and sameness. Matter is at its core conservative, holding to its old ways. Extended, solid, inert, gravitational, *and* change-averse—that is the essence of the mysterious stuff we call "matter" (see chapter 2 on the mystery of matter). Matter will only change when forced to; otherwise it clings to the properties it already has. Maybe in another possible world the stuff that composes objects is naturally volatile, not requiring any new force to change its ways (Aristotelian or Heraclitean matter); but our familiar old matter is a stick in the mud (and I mean this in the nicest way). The law of inertia thus reflects the inner nature of matter as we have it (not that we know what this is).

constancy": simply put, objects remain constant in their properties unless acted on by an outside cause. In other words, constancy of properties is the norm, the default condition, unless the object is subject to outside influence (note that we have to be careful about "outside," as noted above). In yet other words, change is always extrinsically caused (relative to what changes). And more grandly: the universe remains constant in its properties unless there are interactions between objects—unless forces impress themselves on things (from either the outside or the inside). The universe has a deep-seated tendency to remain the same, and it changes only when forces make it change. Given that general principle, we can deduce Newton's first law as a special case—in which case it springs from a far more general and metaphysical-sounding principle (and not from ordinary empirical observation of moving objects).

Indeed, it can rightly claim a kind a priori status. The elemental intuition is that constancies are natural and need no special explanation, while change always requires some cause and explanation. Changes always happen for a reason, but constancy is just nature being its naturally conservative self. Nature is constitutionally inert, we might say, and this is why the law of inertia holds. To put it anthropomorphically: moving objects always prefer to keep on doing what they are doing, from a kind of cosmic dislike of change, and require pressure from outside to behave differently. To slow down (or speed up) would require a change, and change requires energy, and nature just can't be bothered. It is a kind of principle of least action: better to remain the same than undertake anything new, thus minimizing expenditure of energy—though there is no avoiding those pesky outside forces that insist on disturbing one's tranquility. An object by itself is a stick in the mud, roused to take a different path only by interactions with other objects. The law of inertia reflects nature's constitutional passivity—its disposition not to change its ways without due cause.[7]

It might now be complained that there is an important asymmetry between the case of motion constancy and the other kinds of constancy I have cited. It is indeed plausible that change of mass, shape, and color needs an outside cause, and that their constancy is a default state needing neither cause nor explanation, just because such constancies do not involve any kind of change. But in the case of uniform motion, sameness of motion is actually *change of position*, so the object *is* changing over time as it moves at a uniform rate—and this change belies the claim that uniform motion is a case of ontological constancy. If the object is changing position, then here is a

---

[7] I am well aware of how much in this paper I repeat the same underlying thought in different words. I do so with malice aforethought, from self-conscious philosophical methodology. I believe that philosophical illumination often comes from repeated elucidation: one needs to grasp the point from many different angles, so that it sinks completely in. One needs to linger over the point, trying to find the best words, or new words. Mere acceptance of the point is not enough; one needs to be saturated by it. This is why paraphrase is so important in philosophy. This is also why at different periods philosophers keep restating the same points in different vocabulary. Repetition is good (if done right).

change that requires no outside force to explain it—so the operative principle can't be that constancies need no explanation. True, the state of motion does not itself change, but the object clearly changes location as it moves, and yet this doesn't require any cause or explanation. Constancy of shape, say, does not involve change of anything else, but constancy of motion does involve change of position, so the cases are crucially dissimilar—in which case we cannot assimilate the one to the other. The case of motion is so interesting and singular, it may be said, precisely because it involves a change that requires no force to bring it about, i.e., change of position.

There are two replies to this objection. The first is that the state of motion is still a constant, and *it* is what is preserved in unimpeded motion: so we can persist in saying that the law is an instance of the principle that constancies don't require explanation. The property of moving at a particular uniform rate is what is preserved as the object changes places. An object will keep having that property unless an outside force steps in to remove it. But second, change of position is a very special kind of change: it is a relational (or "Cambridge") change. The object itself is unaltered by mere change of position. If an object changed in its *intrinsic* properties over time, and yet no outside (or additional) force were applied, that would indeed be extremely odd (let's say, impossible)—it would imply a kind of spontaneous and groundless inner transformation. That would be really quite different from the case of mass, shape, and color, and not be subsumable under that rubric.

But change of position is not an intrinsic change; it is merely a change of relation to regions of space. Aristotle would explain change of position at uniform speed in terms of a continually renewed force, assuming that even such merely relational change needs a force to bring it about. But Newton is saying that change of position need not be sustained by any continuously applied force, so that *some* changes occur naturally, without any new infusion of force—those that involve merely relational change. This is not the bare claim that constancies don't need explanations but changes do, since some changes apparently don't need explanations either.

However, what Newton is saying is *not* that some *intrinsic* changes can occur without explanation—which, as I say, would be quite remarkable. Rather, the changes in question are merely (so called) relational changes: the object doesn't really *change* as it shifts position—it simply alters its relation to space and other objects. The object remains the same, intrinsically speaking. So we can reformulate the general law to assert that all *intrinsic* change requires an outside force, and the case of uniform motion does not involve intrinsic change. The case is really no different from the following: an object varies its spatial relation to other objects as *they* move, but this kind of "change" requires no external force acting on the first object to explain it. This would clearly not be a counterexample to the claim that all intrinsic change needs a force to explain it. So the law of inertia, which speaks of changing locations over time, is not a counterexample to the claim that nothing can intrinsically change unless a force brings this about. We can thus still say that the underlying principle of the first law is that constancies need no sustaining force while (intrinsic) changes always need the imposition of a force.

I said earlier that the constancy of rest has been found less surprising than the constancy of uniform motion. People feel that objects naturally stay still, while continuing motion requires some sort of effort. But it is not clear that this difference of attitude can survive discounting the effects of friction and gravity. On earth a stationary object is subject to the forces of friction and gravity, as well as having its own intrinsic inertia: what makes it stay put is a combination of all three factors. The same is true of the observed motions of objects on earth. But once we idealize away from friction and gravity, rest can seem just as precarious as continued uniform motion: maybe it does require an extra force to keep a thing rooted to the spot, once friction and gravity are discounted—so Aristotle might think. Really, the case of rest is just like that of free motion: both are cases in which only the inertia of the body itself prevents a change of motion. Spontaneously shifting from a position of rest begins to look just as feasible as spontaneously changing the state of uniform motion—neither more nor less. It is only focusing on the terrestrial case that fosters the illusion that rest is somehow inherently more stable than continued uniform motion. In reality, both arise equally from the self-same principle, namely that objects maintain their state of motion—zero or nonzero—unless something comes along to deflect them. Neither claim is more self-evident than the other, properly considered. It is no more certain or intelligible that an object could not start to move of its own accord in empty space than that it could not change its direction or speed of motion of its own accord. Both involve departures from the status quo, and that is what Newton's first law proscribes.

In addition, from the point of view of relative motion, there cannot be a deep difference between the two cases. Newton himself believed in absolute motion, so there is certainly for him a basic difference between rest and motion: but for those who recognize only relative motion there cannot be. All objects are at rest relative to themselves, according to that view; and all move relative to other objects. So a case of not moving from a position of rest *is* a case of retaining uniform motion—relative to *some* chosen rest-frame. No state of motion is inherently a case of rest or of uniform motion, on the relative view; so the preservation of either amounts to the same. And even for those who believe in absolute motion, it has to be admitted that which objects are absolutely at rest and which absolutely moving cannot be inferred from observation, since only relative motion can be observed. We can't *tell*, of any given object, whether a change of its motion is a case of shifting from a state of absolute rest or a change in the absolute motion already in progress. Again, the difference between rest and motion is not as pronounced as one might suppose. So, if the analysis of the rest case I have given is plausible, so must the like analysis of the uniform motion case: both are cases of natural constancies that require no prodding to keep them in being.

The inertia law says that a freely moving object moves eternally in a straight line—not a circular or elliptical one. Aristotle held that celestial movements are naturally circular; Newton explained the elliptical orbits of the planets in terms of the outside force of gravity (plus inertia), not the inherent nature of planetary motion (without gravity the earth would sheer off from its orbit around the sun and begin an eternal rectilinear journey across the universe, unless subject to some new force). But *could* there be a universe in which the law of inertia implied persevering circular motions instead of straight lines? Why couldn't it be the case that objects in free motion naturally describe circular motions or even some sort of zigzag motion? Why the straight lines? The answer is that only straight lines involve no change of direction. If an object undertook a circular path, it would be changing direction from moment to moment. Only traveling in a straight line involves no shift of direction; so a circular path would involve *changes* of direction without an outside force (this is even clearer for a zigzag path). A change of direction would require an outside force precisely because it is a genuine change in the state of motion of the object—and that is ruled out by the law of inertia. Circular motions are ruled out by the law of inertia as much as accelerated motions: *these* don't persist without outside forces, because they involve changes in the state of motion, either in respect of direction or speed. Only uniform motion in a straight line persists without external support—not accelerated motion in a circle.

This confirms the analysis in terms of constancy and change, and what requires explanation and what does not. A straight line means there is no change of direction, and that is exactly what the analysis in terms of persistent constancy requires—no change, no explanation needed. Also, since a straight line is the shortest distance between two points, the law of inertia tells us that motion naturally occurs in the most economical way: nature prefers objects to reach new places by the shortest path possible. This is an aspect of that conservativeness in nature that I earlier mentioned. To persist in a circular motion would be for nature to allow objects to reach new destinations by circuitous routes—and that is a waste of time and effort. Straight lines satisfy nature's craving for economy.

The difference between Newton and Aristotle can be put in terms of what does and does not need explanation (where this involves the citation of a force). For Newton, constant uniform motion in a straight line requires no explanation (no continuously applied force), while change of motion does, either of direction or speed. For Aristotle, uniform motion does require explanation (a continuously applied force), while change of motion does not—it is just the initial force wearing itself out over time. Slowing down needs no positive applied force to counteract motion in progress, for Aristotle, and hence is natural and in need of no explanation; whereas for Newton slowing down is a positive event needing outside coercion. In effect, Aristotle takes change of motion (of some kinds, at least) not to need explanation, while constancy of motion does; and Newton takes the opposite position. It is a question of what can be taken for granted and what cannot: of what is natural and what contrived, of the spontaneous and the adventitious. It is a question of what things do of their own accord, in conformity with their inner nature, versus what external forces visit on them. We could put it by saying that Newton taught us that objects in motion are more naturally active than Aristotle supposed and ordinary observation suggests: their natural state is not immobility but eternal activity. If Aristotle thought the natural goal of matter was to come to rest, Newton thought that matter had a natural tendency to keep moving—eternally. Motion is more in the nature of matter for Newton than for Aristotle. Newton's first law of motion thus changes the philosophical *Gestalt* considerably, simple as it sounds.

What is the larger significance of the shift that Newton (aided by Galileo and Descartes) brought about? I think it is that Newton put ontological constancy at the heart of nature—not change, but permanence. He made the changeless and eternal the basic principle of the physical world: this is what reality naturally is, when forces do not intrude to subvert it. Things naturally stay the same, even active things like motion; they don't change of their own accord. In this he sides with Parmenides and Plato over Heraclitus. On the Heraclitean view, nature is essentially in flux: things change spontaneously, of their own nature, without external prompting. Nothing stays the same: being is necessarily becoming. What is here today is gone tomorrow (or an instant later): nothing is preserved. For Aristotle, motion is essentially a phenomenon of change: change is inherent to it, part of its essence. It no sooner starts to exist than it spontaneously dissipates, unless something steps in to prop it up and keep it going. But Newton sees the continuing order beneath the flux: motion unimpeded is eternal and uniform. Its nature is to stay the same: it changes only because of extraneous agencies. Interaction is the source of change in motion, not motion considered in itself.

As I have suggested, this can be seen as a consequence of a more general principle of changelessness, so that the universe is conceived as fundamentally unchanging—that is, prior to interactions between its constituents. It is as if the objects and processes in the universe individually want to remain constant but must contend with unruly interacting agencies. However, for Heraclitus and his followers the universe exists in an inherent frenzy of upheaval and inconstancy: things change of their own volition, as it were. Isolation protects against change for Newton, while isolation is no refuge from change for Heraclitus. Growth and decay are fundamental and constitutive on the Heraclitus-Aristotle view, but they are superficial and incidental on the Newtonian view (as with Parmenides and Plato). Here we see one of the largest and most ancient of metaphysical divides centering on a basic law of physical science—which is why I have spent so much time discussing it.

The foregoing has a bearing on aging and death, as well as other processes of decline. It is commonly assumed that these are natural results of the pure passage of time—the inevitable deterioration and extinction of the body. The body simply wears out, as motion wears out for Aristotle. Eternal life would require some kind of supernatural intervention, the application of a divine force: it is not the natural lot of human existence, or animals and plants. But from a Newtonian perspective this would be the wrong way to look at it—and it is wrong according to the best current science of the aging process, too. Instead, aging and death result from active forces impressed on the living organism. Organisms would go on living forever unless forces undid them; they are not inherently or constitutionally mortal, at least in one special sense. Death is not an inevitable consequence of life; it is not built into life as such. If all interfering forces (internal and external) could be removed from a living organism, it could live forever—just as a moving object will continue to move for all eternity if not interfered with. Once something exists it will continue in existence until some external force comes along to extinguish it—motion, mass, shape, color, life. Consciousness itself would continue forever, if only nothing interfered with its existence.[8] Things and their properties naturally persist, considered purely in themselves; they don't naturally perish. That is the law of inertia, suitably generalized. Everything has a tendency to persist in its state of being, metaphysically speaking, including conscious organisms, unless acted upon by inimical forces.

Of course, nothing can come to be—nothing can attain a given state of existence—without some process of change to produce it, so that change is required for the onset of existence: but once a state of being is attained nothing is required for it to persevere—unless it is to stave off outside destructive forces. Immortality is implicit in every state of being, considered in itself, though change is generally required to bring that state into existence. Things cease to be, not because of what they are in themselves, but because of forces beyond them. Objects continue to instantiate properties unless something active occurs to remove the property. That is in the very nature of instantiation.

The law of inertia, thus generalized, says that if it were not for outside influence everything would be immortal—not just uniform motion in a straight line but every other property of things. It is never internal to any attribute that it should spontaneously cease to be instantiated. All change of properties is imposed from outside those properties. Complete isolation would thus ensure universal perseverance, if only it

---

[8] If consciousness is inherently a process of change, however, it does require the continuous action of a force to keep it going, because then for consciousness to exist *is* for it to change. The persistence conditions of consciousness would thus be like those for war: a war will only persist if the world keeps changing in suitable ways, and hence new forces are needed to make wars persevere; wars are not self-propelling (consider "once at war always at war, unless prevented by an outside force"). So, strictly, assuming a like view of consciousness, we would only get a static time-slice of consciousness persisting to eternity in the absence of disruptive outside forces—as it might be, a specific visual experience possessed at time t. Suppose the brain is in configuration C at time t, and suppose that C is the realization of a conscious state S at t: then if C continues in existence indefinitely because of a lack of external disruption, S will likewise continue in existence indefinitely. If the brain is stable over time without external interference, then so is consciousness, assuming that certain brain states are sufficient for conscious states. Likewise, if we freeze a war at a given time, then the total state will persist unless interfered with by an outside force—though the war itself will be over (because a war is a process of change). If we freeze consciousness at a given time, then the state of consciousness at that time will persist without external interference—though consciousness itself may be over (on the assumption that "consciousness" refers to a process of change). (Much the same can be said about the persistence of life.)

were possible. The law of inertia, when generalized, asserts a fundamental permanence at the core of things. Eternity is written into every fact.[9]

Let it be noted, finally, that the law of inertia, since it concerns motion, inherits the epistemological obscurity of motion, as discussed in the previous two chapters. If real motion is captured neither by relative motion nor absolute motion, then the uniform motion that Newton asserts to persist is neither of these. Adopting the structuralist we-know-not-what position gives the result that the property that persists according to the law of inertia is a mysterious property whose nature we fail to grasp. We know that, whatever it is, it is preserved, but we have no clear and distinct idea of what it consists in—over and beyond its mathematical properties and its effects on our senses. The law is thus a self-evident and a priori truth about something inherently mysterious to us.

---

[9] Being written in doesn't mean that it cannot be erased—but the erasure requires something from outside the fact to make it so. Left to itself the fact would persist indefinitely. As we might say, every (nonprocess) fact is "locally eternal." Thus Newton's law of inertia (to restate it one last time) says that uniform motion in a straight line is *locally eternal*.

6 Mass, Gravity, and Motion

Mass is connected to motion in two ways. First, gravity is proportional to mass, and gravity determines the motion of bodies within the gravitational field: the greater the mass, the greater the gravity—and hence the greater the motion caused in affected bodies. The pull of gravity is directly reflected in the motion of the object acted upon. Second, mass determines inertia—the amount of force it takes to move an object from rest or from a state of uniform motion. The more massive a body is, the harder it is to impart motion to it. So mass determines both gravitational force and inertia, and both are defined in terms of motion.

Of these two, the second is more intuitive. The more matter an object contains the harder it will be to move it, since there is more inertia to overcome the more mass there is: if each part has a certain degree of inertia, then surely the sum of those inertias will be proportionately greater the more parts there are. To move a given particle we have (generally) to move all the other particles cohering with it, and this will be harder if there are more particles to move. But the law of gravity is not intuitively self-evident: how can an increase of mass in a given object make *other* objects, at arbitrary distances, move differently? Why, indeed, should the mass of one object determine another's movements at all? Volume doesn't, or shape, so why should mere quantity of matter have such a dramatic effect? (So my question is not answered by General Relativity: mass causes the bending of space, but why is *that*—couldn't volume or shape have done the same?)

Inertia self-evidently depends on mass, but gravity's dependence on mass seems arbitrary. Why should adding to the mass of an object over *here* make any difference to the motion of an object over *there*? This is not Newton's original worry about action at a distance through a vacuum or the occult nature of gravitational attraction; it is the more basic concern that one magnitude seems to have nothing intrinsically to do with the other. Maybe the *motion* of one object could affect the motion of another through some sort of cross-spatial force (agitation in one place produces agitation in another), but how could the mere *bulk* of an object have such an effect? How can mass, a static measure of sheer material quantity, bring about the motion of distant objects? The effect seems unrelated to the cause. The cause is heaviness, the effect is accelerated motion: what has the former got to do with the latter? When an object accelerates to earth under the influence of gravity, it undergoes severe changes of motion, yet the mass of the earth remains static and unchanging (it is the distance that changes). Simply put: how can the mere amount of matter in an object exert any effect on the motion of distinct and remote objects?

To answer this we would presumably need a better understanding of the nature of matter, but as I have argued elsewhere we don't know what matter is.[1] However, my concern here is with another question: the specific connection between inertia and gravity. The relationship is this: the greater the inertia of a body the greater its gravitational force. This means that the harder it is to move a body the more movement it causes in other bodies. This should strike us as strange: if a body in motion collides with another body its motion will be retarded, because of the inertia of the second body, proportionately to its mass. But that very mass will make other objects move from their prior paths, causing them to accelerate as they enter the object's gravita-

---

[1] See chapter 2. Newton would appear to agree, writing: "Hitherto we have explained the phenomena of the heavens and of our sea by the power of gravity, but have not yet assigned the cause of this power. This is certain, that it must proceed from a cause that penetrates to the very centres of the sun and planets, without suffering the least diminution of its force; that operates not according to the quantity of the surfaces of the particles upon which it acts (as mechanical causes use to do), but according to the quantity of the solid matter which they contain, and propagates its virtues on all sides to immense distances, decreasing always in the duplicate proportion of the distances.... But hitherto I have not been able to discover the cause of those properties of gravity from phenomena, and I frame no hypotheses; for whatever is not deduced from the phenomena is to be called an hypothesis; and hypotheses, whether metaphysical or physical, whether of occult qualities or mechanical, have no place in experimental philosophy" (*Principia*, 442–43). Later philosophers of science have often quoted the last few clauses of this passage (following "for"), taking them to express an antipathy towards ultimate causes and a foreshadowing of positivism. But the preceding sentences make it clear that this is by means Newton's message. He expresses no doubt that gravity has an objective cause in the matter that composes bodies—the power must have a "categorical ground," as we say today. He simply proclaims his ignorance of this real cause, and acknowledges that without knowledge of it his account of gravity is incomplete. He "frames no hypotheses" not because he disbelieves in ultimate causes but because he cannot see how to reach the cause of gravity from the phenomena that manifest it; and if causes cannot be deduced from phenomena, they have no place in empirical science ("experimental philosophy"). Not that they don't *exist*: but their nature can only be a matter of speculation—which is a purely epistemological point. I think it is quite clear from this passage that Newton believes that the secret of gravity resides in the hidden nature of matter (and just before the passage quoted he expresses his belief that the "inward substances" of bodies are unknown to us).

tional field. If an object hits the earth, it slows down; but if it enters the earth's gravitational field, it speeds up. The same mass causes deceleration and acceleration in other objects. The less motion an object itself contains, as measured by inertia, the more motion it produces, by way of gravity. The sun has less of a tendency to move than the earth, because of its greater inertia, but the sun *produces* more motion than the earth. In other words, the tendency of an object to move is inversely proportional to the amount of movement it generates: the less prone to move an object is, the more motion it produces in other objects. As I say, this seems odd. The law of gravity is telling us that if you add to an object's inertia (by increasing its mass) you will make it bring about more movement elsewhere: if you slow it down, you will speed other things up. By reducing its propensity to motion you ensure that it imparts more motion to other things. It is as if what it loses in motion the other object gains.

Can we make this any more intelligible? Or is it just a brute fact about our universe that we have discovered empirically to be so? My hypothesis is this: the total motion of any system of gravitationally interacting bodies is a constant. Take two bodies of equal mass at a certain distance in empty space: both have the same inertia and gravitational force. Motion will occur between them: other things being equal, they will fall into each other at the same rate, meeting midway between their two original positions. Let us say that the total resulting quantity of motion is m, which sums up both distance traveled and velocity. Now suppose we take half the mass of one and add it to the other. We thereby double the inertia of the second body and also double its gravitational pull, while reducing by half these two magnitudes for the first body. But the amount of motion that results will be the same, since the loss of motion due to inertia on the part of the more massive body will be compensated by the increase of motion in the smaller body acted upon gravitationally (it now has less inertia than before): the total will still be m. The bigger object will travel less far and more slowly than before because of its increased inertia, but the smaller object will travel further and faster because of its decrease of inertia.

If this is right, then the law of gravity, combined with the law of inertia, is saying something interesting and elegant: in any gravitational interaction the quantity of motion is a constant (*modulo* constancy of total mass). As an object increases in mass, it loses in motion of its own, but increases the motion of other objects. There is a kind of conservation of motion at work. A black hole, say, will cause tremendous motion in surrounding objects, as it sucks them rapidly in, but it is proportionately immobile in itself—what it gains in motive power it loses in move-ability. Similarly, a very low-mass object, like a mote of dust, will have very little gravitational force, but also very low inertia—easy to move but incapable of doing much remote moving.

If we consider the universe before matter began to clump into stars and planets and so on—when it was a relatively homogeneous gas—it may look as if movement was less at that time than later, when big dense bodies started to pull things around. We can imagine particles of matter held in equilibrium in a gaseous state without the kinds of gross orbits and falls that we observe today.[2] Thus it may look as if motion has increased over time, despite the constancy of total mass. But the point I am making is that the increased large-scale motion that resulted from clumping, combined with gravity, corresponds to a decrease in motion (at least dispositionally) on the part of the particles that formed the clumps. More massive bodies have greater inertia, so the particles that make them up are less prone to move than when existing in a free state. Given the connection between gravity and inertia, via mass, it is not possible to increase the former without increasing the latter—and this means that the move-ability of matter is reduced when its power to move other objects is enhanced. I don't mean that the *actual* movement of a piece of matter is necessarily decreased when its gravity is increased, since it may be continuously at rest; I mean that its *propensity* to move is decreased. The movement *potential* of matter is what remains constant under increase of mass. When a particle is inside a more massive body it is prone to move less than when it is part of a less massive body or when it is in a free state, because of the inertia of the whole, but the containing body produces a correspondingly greater amount of motion in other bodies by virtue of gravity. To produce more motion in something else a body has to become more sluggish in itself.

It might be objected that everything I have said so far assumes that mass should be defined in terms of inertia, i.e., the tendency of an object not to be moved: but couldn't we define mass in terms of an object's ability to move *other objects* (instead of being moved by them)? Then we could say that the greater the mass the more force the object can exert on other objects, and hence the easier it is for it to move them. If an object leans on another object, it will induce motion proportionately to its mass—that is, it will overcome the inertia of the other object to varying degrees. I see no principled objection to such a definition of mass; the question is whether it changes the picture I have been presenting. For now we can say that an increase of mass actually *increases* total movement, in virtue of the greater propensity of the object to cause motion in other objects (by means of contact).

So, as we increase gravity we also increase motion induced by physical contact; there is then a net increase of motion. Suppose an object doubles in mass: then it doubles in gravity and in addition it doubles in its capacity to move other objects by collision or other types of contact. In other words, motion caused by action at a distance and motion caused by contact both increase when mass increases; the two are directly proportionate, not inversely. The very thing that makes objects harder to move—increased mass—also makes it easier for them to move other objects by contact. Indeed, one might conjecture that the two cancel out: the total (local) state of motion is the same after the increase of mass as before, since what is lost in the way of inertia is gained in the way of power to move other objects by contact.

---

[2] Or the motions might be slight for each particle of gas, though the total motion may be very large because of the number of particles—we just wouldn't *see* much motion, on account of the particles being so minute. We won't observe the kinds of spectacular motions that occur when matter clumps into bodies of very large mass, such as stars. Still, the quantity of motion could be conserved.

If this is so, then increases of mass do increase total movement, since they increase movement due to gravity, as well as increasing motive power by contact. However, the increase of contact-based motive power due to mass is correlated with a loss of mobility due to inertia. All these types of motion are interrelated: more motive power, less mobility, but more gravity; more gravity, less mobility, but more motive power; greater mobility, less motive power, and less gravity. Mass is that which increases motive power by contact, and increases motion due to gravity, but decreases motion due to inertia. All three types of motion are functions of each other. Given information about one type of motion, it is possible to infer information about the other two. For example, if you know the mobility of an object (i.e., its degree of inertia), you can infer both its motive power by contact and its gravitational force.

The main point I have wanted to get across is the interrelationship of two sorts of movement—that of bodies exerting gravity and that of bodies acted upon by gravity. These are not independent, because mass is responsible for both of them. There is the movement at a distance that is caused by gravity, and there is the movement by contact that expresses inertia (of course, inertia is also manifested in an object's resistance to gravitational pull). I have tried to articulate these interrelations, so that the full meaning of the law of gravity can be appreciated.

To recapitulate: the easier it is for a body to move another by contact, the easier it is to move another by gravity; and the harder it is to move a body by contact (or by gravity), the easier it is for that body to move another by gravity. What we have here is a systematic interrelationship between sets of movements, mediated by mass. The general and basic truth expressed by the law of gravity is therefore that one sort of movement is related by law with another. We can thus predict the movement of remote objects, using the law, from information about the local movement properties of a given object (i.e., from its inertia). If you know that an object is very hard to move, you know that it can easily move other remote objects; and there is a precise mathematical relationship here. We might put the inverse square law by saying that the gravitational effects of inertia dissipate exponentially with distance. The further things are away, the less the inertia of a given body can accelerate their motion. The sluggish behavior of one body has less influence on the agitated behavior of another body the further the two bodies are apart. Thus the sun's relative lack of movement can be used to predict the earth's profligacy of movement once we know the distance between them. The law of gravity states an inverse relationship between the respective movements of two bodies, as well as an inverse relationship between force and distance. The less a thing is prone to move by impressed forces (of whatever kind), the more it is prone to induce movement in other things remote from it.[3]

---

[3] My methodology in this paper has been one of elucidation by reformulation (see also chapter 5). I endeavor to spell out the law of gravity in such a way that its full implications are made clear. I doubt that I am saying anything that will be new to a physicist, but as a philosopher I find it illuminating to restate the content of the law so as to make its consequences for the total state of motion explicit. I suppose this is a kind of conceptual analysis: articulating the full import of a proposition. Such an exercise can be highly nontrivial.

7 Electric Charge: A Case Study

Hume described causation as the cement of the universe. The metaphor, though famous, is not altogether apt. Cement makes bricks stick together in a stable and static whole, while causation makes things change and flow: it is the oil of the universe. And cement is a visible and tangible substance—we have "impressions" of cement—quite unlike the relation of causation, as Hume understands it. But electric charge has a better title to that description: it is the principle of cohesion in material things, what holds particles together in stable wholes. In particular, the attractive force between electrons and protons is what keeps the atom stable and allows chemical bonding to occur. Electricity is not an isolated phenomenon, constituting lightning and powering human devices, but runs through everything. Ordinary objects are electrically neutral, but that is only because the charged particles that make them up are in a state of equilibrium; in fact, every object is throbbing with electric charge. Not surprisingly, then, the study of electricity, and with it magnetism, is now a large part of physics, and there is a very advanced theory detailing its laws. We learn, for example, that charge is quantized and obeys a conservation principle—as well as Coulomb's law, Gauss's law, and Maxwell's equations; and of course there are enormous practical applications of the theory. We know a lot about how charged particles behave, and hence how matter operates by means of electricity. It would be nice, then, to know what electric charge *is*. What is it to be charged, exactly? When a body goes from being neutral to being charged, what is the nature of the change it undergoes? What is the state it ends up in? When a body goes from being circular to being elliptical, we know what the change consists in—it is a matter of simple geometry—but what is it to come to possess an electric charge? What *is* electricity?

We naturally turn to the physics textbooks. The chapter called "Electric Charge and Electric Field" in Young and Freedman's standard 1700-page textbook *University Physics* begins thus:

Electromagnetic interactions involve particles that have a property called *electric charge*, an attribute that is as fundamental as mass. Just as objects with mass are accelerated by gravitational forces, so electrically charged objects are accelerated by electric forces. The annoying electric spark you feel when you scuff your shoes across a carpet and then reach for the doorknob is due to charged particles leaping between your finger and the doorknob. (A lightning bolt is a similar phenomenon on a far grander scale.) Electric currents, such as those in a flashlight, a portable CD player, or a television are simply [!] streams of charged particles flowing within wires in response to electric forces. Even the forces that hold atoms together to form solid matter, and that keep the atoms of solid objects from passing through each other [is that what impenetrability is?], are fundamentally due to electric interactions between the charged particles within atoms. (792)[1]

Notice that we are immediately introduced to the notion of "electromagnetic interactions" without any prior account of what kind of interactions these might be—so far, these are just interactions with a fancy name. We then learn that forces resulting from something called "electric charge" accelerate objects and that this is a fundamental feature of objects. In other words, certain interactions, involving certain motions of particles, are designated "electrical," and then we introduce an attribute called "electric charge" to be the origin of the force that accounts for these interactive motions.

We might naturally ask: what exactly is this thing called "electricity" and how does it give rise to the force associated with something called "charge"? Is it just a *name* for whatever it is that is responsible for the interactions we observe, with no descriptive content? A force is evidently at work, which we choose to *call* "electrical," and then go on to postulate that the force is borne by an attribute or quantity

---

[1] Hugh D. Young and Roger A. Freedman, 11th ed. (Pearson, Addison Wesley: New York, 2004). As I parenthetically note, the authors are casually confident that material impenetrability is the result of repulsive electric forces, which is highly questionable: see chapter 1. Do they think that neutrons, which lack charge, are therefore entirely penetrable? Do they think that with enough force one particle of charged material could overlap with another—you just have to accelerate them fast enough? Is a particle with a greater charge more impenetrable than one with lesser charge? If the charge is very low, do they think that the particle is easily penetrated? Is the reason for the penetrability of space that it *lacks* electric charge? And what result in physics do they think establishes that the impenetrability of matter just consists in repulsive electric charge? What about mutually attractive particles? Isn't this really just a piece of (dubious) metaphysics on their part? We can grant that charge can operate to keep particles apart by means of repulsion, but is that the *only* thing about matter that is relevant to its impenetrability? What about the basic fact that it *occupies* space?

we choose to label "charge." We could have chosen, in the same spirit, to speak of "gravitational charge," meaning the attribute that is responsible for the gravitational force we observe—and then gone on to discover that this attribute is actually mass. But so far we just have labels that sum up the interactions we have observed: the interactions are driven by a force we decide to call "electrical" and we suppose that this force is based in a property we decide to call "charge." These two notions have no content, as so far explained, besides the observations we use them to summarize. The idea of "electromagnetic interactions," construed as a basis for introducing the ideas of electricity and charge, is motivated purely by the observation of motions not explicable in terms of forces we have already recognized, such as gravity. The concept of charge is *not* defined by something like a certain arrangement of particles in space or the shape of the underlying particles or the agitation of particles or some such categorical property. Charge is understood solely as that which produces a certain distinctive pattern of acceleration. We observe the pattern, notice it is not subsumable under forces already identified, and then posit a new quantity presumed to underlie the manifestations of the new force.

So far, then, the concept of charge is defined purely in terms of its effects, i.e., distinctive types of motion. Perhaps aware of the thinness of what has so far been offered, Young and Freedman go on to relate the concept of electricity to experiences assumed to be familiar to their student readers: sparks of "static," bolts of lightning, and the "current" coursing through everyday electrical devices ("Ah, now I understand!"). It is not that these give any real insight into the nature of electricity, but they at least relate the theoretical concept introduced to experiences assumed to be widely shared. Then the authors revert to talk of charged particles and electrical force, with no further explication of exactly what these things consist in. They are really, in this passage, introducing the reader to the way physicists *talk*. The only hard reality in any of this is the specific type of motion associated with this thing called "electricity." The rest is mere labeling.

The textbook goes on to explain and elucidate the basic concepts of the theory of electromagnetism as follows: "The ancient Greeks discovered as early as 600 B.C. that when they rubbed amber with wool, the amber could then attract other objects. Today we say that the amber has acquired a net **electric charge**, or has become *charged*. The word 'electric' is derived from the Greek word *elektron*, meaning amber. When you scuff your shoes across a nylon carpet, you become electrically charged, and you can charge a comb by passing it through dry hair" (793, bold emphasis in the original). This passage also is instructive (for its lack of instruction). Note how the authors cagily write, "we say" the object has become electrically charged, as if to acknowledge a stipulation, not a description of what the attribute in question is. Also, we learn that the word "electricity" simply derives from the word for the stuff (amber) in which the Greeks first noticed something funny going on, an odd kind of attraction when that stuff was rubbed with wool: so there is no illuminating etymological content here to fill in the picture.[2] And then we again have the attempt to relate the concept to something students might be familiar with, in the line of shoes and combs, specifying observable effects of the thing we are deciding to designate "electric charge."

The exposition then proceeds to describe the effects of rubbing fur and silk on plastic and glass rods and noting the changes in the attractive and repulsive properties of these objects: some of the items attract others, while those same items will repel yet others (charged glass rods repel each other, while charged glass and plastic rods attract each other). So, we observe that the rubbing actions produce both attractive and repulsive effects, and we then suppose that becoming "charged" involves acquiring a disposition or power to produce such varying effects. The authors conclude: "These experiments and many others like them have shown that there are exactly two kinds of electric charge: the kind on the plastic rod rubbed with fur and the kind on the glass rod rubbed with silk. Benjamin Franklin (1706–1790) suggested calling these two kinds of charge *negative* and *positive,* respectively, and these names are still used. The plastic rod and silk have negative charge; the glass rod and the fur have positive charge. **Two positive charges or two negative charges repel each other. A positive charge and a negative charge attract each other**" (794). That ends the conceptual explanations—the ontology of electricity. Thereafter, it is all equations, laws and calculations, along with the extension of these notions to elementary particles.

Again, the wording is telling, if not altogether forthright: having made the relevant observations, Franklin simply *stipulated* that we call the alleged duality of charges "positive" and "negative," and these "names" (not descriptions!) are still used. He could have made the converse stipulation, in which case we would now describe the electron as having positive charge and the proton as having negative charge. The labels are entirely conventional. He could equally have called the charge on the plastic rod rubbed with fur the "red charge" and the charge on the glass rod rubbed with silk the "blue charge." Then we would have said: "two red charges or two blue charges repel, but a red and a blue charge attract" or "like colors repel and unlike colors attract." Or, perhaps more revealingly, he could have invented two new words or symbols that function purely as labels, with no suggested descriptive content. The dichotomy of "negative" and "positive" is merely verbal. The epistemological situation here is best captured by a declaration of this form: "There's a force here I am calling blah-blah [=electricity] and it comes in two types, which I am labeling A and B [=positive charge and negative charge]." What prompts the whole apparatus is the observation of attraction and repulsion on the part of the

---

[2] Imagine if the English first discovered electricity while messing around with wool, calling the new force "wool." The textbooks would now have long chapters entitled WOOL and would speak routinely of "woolly interactions." Would we then be quite so inclined to think we knew what electricity really is? Wouldn't we want to ask, petulantly, what is this thing called *wool*?

same items in relation to other items: of that we do have a clear descriptive conception.[3]

It is easy to run away with the idea that talk of positive and negative charge implies some sort of physical reality, namely that a negative charge is the *absence* of some quantity while a positive charge is its *presence*. Some writers even speak of "negative electricity" and "positive electricity." But this is obviously quite wrong, since both types of charge are "positive" in the sense that they both involve (nonzero) amounts of energy: so-called negative charges are just as potent and substantial as positive ones (just as red ones are as potent as blue ones). It is not that the two types of charge sum to zero in the sense that minus two apples added to plus two apples gives zero apples; rather, the two "positive" charges (negative and positive) reach an equilibrium—they push and pull in such a way that nothing happens. It isn't that a neutral object like a table has *no* electrical energy in it, that it is electrically dead: it has a lot, but the forces are opposed to each other. It is like one hand pushing another with equal force, not like two hands lying limply side-by-side. A negatively charged particle can attract and repel, depending on what it is paired with, and so can a positively charged particle: the physical reality is indistinguishable. It isn't that so-called positive and negative charges are different *physical* kinds of charge: they both perform the same attractive/repulsive double act, only choosing different partners. Both "kinds" of charge produce accelerations of the same type, so they don't differ in their intrinsic physical nature. Labeling them "positive" and "negative" is simply a convention we use to keep track of the varying interactions we observe, which can be either attractive or repulsive. That is where the duality comes from, not from the physical nature of the two types of charge.

It is also conventional that we declare that like charges repel and unlike charges attract: we could have said the opposite consistently with the facts. Why *not* say of mutually repelling particles that they have unlike charges? They evidently don't have an affinity for each other, so isn't it more natural to describe them as unlike? And mutually attractive particles might naturally be described as having like charges, on account of their affinity. The familiar statement "like charges repel and unlike ones attract" is not an empirical discovery but a linguistic stipulation. It is not that we

---

[3] At least we do if we grasp the nature of motion: but see chapter 3. What we really know is that the motions (whatever they are) dubbed "electrical" are not the result of the force we dub "gravitational," because that force cannot produce those kinds of motions; so we need to postulate a new kind of cause for the former type of motion. We know the motions come from different forces, though we don't know what motion intrinsically is or what the forces derive from. Note, too, that we would have had to postulate a greater range of forces had we observed more kinds of effect: not just attraction and repulsion at right-angles to the object's surface, but also attraction and repulsion at different angles or tracing curvilinear as well as rectilinear paths. We would then need several new names for the causes of these various types of motion. It is the type of motion we observe that dictates the forces we attribute to the world (Newton showed that terrestrial and celestial motions are really of the same type, so that we need just one force of gravity to account for them). There is no "direct route" to the forces of nature.

have examined mutually attractive objects and discovered that they differ in their intrinsic attributes, while those that repel have been found to be identical in their attributes—specifically, we haven't empirically discovered that the nature of the charges in attractive pairs is one of intrinsic difference, while it is the same for repelling pairs. The statement is not an empirical law relating particular attributes, but a bookkeeping rule to keep track of the various motions we observe. Having positive charge is not like having mass, while negative charge is the absence of mass (cf. "negative mass")—or some other physical attribute. Having positive charge *or* negative charge is like having mass—a definite physical attribute: but the *difference* is not a real physical difference.[4]

To see what is going on here, we need to step back and remind ourselves of the facts. The gravitational force is always attractive, so we never observe one pair of massive bodies pulled together while another pair is thrust apart. There is a single kind of effect of gravity, and so we can make do with a single force. But with electricity and magnetism the same particle can be both attractive and repulsive, depending on which other particle it is paired with it: electrons and protons repel particles of the same kind as themselves, but they attract particles of the other kind. There is a duality of effects here. We speak of two types of charge because we observe two kinds of motion in respect of a single particle, as it either attracts or repels another particle. If we observed only one kind of motion, say attractive, we would speak only of a single type of charge (as we do with gravity). We can't ask whether a given particle is an attractive or repulsive particle *simpliciter*, since it is always both, depending on what we pair it with. So we don't speak of two forces, attractive or repulsive, which different particles have: we must allow that the same particle has both dispositions. This is where the binary apparatus of positive and negative charges comes from, not from inspection of the internal structure of the particles. By saying that a given particle has one type of charge—positive or negative, red or blue, A or B—we can summarize the fact that it interacts differently with other particles, attracting or repelling. We just need a way to encapsulate the fact that the induced motions are of different types, attractive or repulsive. If gravity behaved similarly, we would speak of two types of "gravitational charge," positive and negative.

The basic fact is that a given particle has two dispositions or powers—to attract and repel. Which disposition gets activated depends on the particle that comes within its field of force. If we sort the particles into two piles, calling one positively charged and the other negatively charged, then we can cover the ground, saying that if like charges come together we get repulsion and if unlike charges come together we get attraction. But there is no more content to this description than the fact that a given particle has the two dispositions I mentioned. Electrons are disposed to repel

---

[4] We could say the difference is relational not intrinsic: electrons relate differently to other electrons from the way they relate to protons, and similarly for protons vis-à-vis other protons and electrons. But electrons and protons don't have different intrinsic physical attributes labeled "negative" and "positive."

other electrons but to attract protons, and similarly with protons *mutatis mutandis*—that is really all there is to it. All the talk of positive and negative charge is just a way to record this basic fact. It has no further descriptive or explanatory content. Above all, we have certainly not specified the *mechanisms* of the (so-called) electric force by all this talk of positive and negative charges: the positive/negative distinction is purely conventional and stipulative, and the talk of "electric charge" is so far just a label for whatever underlies the interactions we have observed.[5]

Contrast gravity and mass with electricity. A body exercises a gravitational force on other bodies, so that it has a disposition or power to induce motion in distant bodies. Let us call that disposition G: it is specified by way of the laws that govern gravity. If we ask what grounds G—what its categorical basis is—we get the familiar answer, viz. mass: the magnitude of gravitational force is proportionate to the mass of the body. And there is an independent criterion for the possession of mass—independent, that is, of its gravitational effects—namely, inertia. So we can say that the body has G in virtue of its mass: mass is the "mechanism" of gravity, the attribute that gives rise to the power. (Of course, this does not remove all questions about gravity, since it is unclear *how* mass operates to produce gravity: but at least we can specify an attribute of a gravitating body that underlies its disposition to produce gravitational motions.)

But in the case of electric charge, there is nothing corresponding to mass. To have a charge just is to have the power to produce certain accelerations in other objects; at any rate, nothing has yet been said about what charge is *apart* from such a power. We can't say that a particle has a charge *in virtue* of some property C that underlies the disposition and from which the disposition follows, as we can in the case of mass and gravity. It was a genuine discovery that mass is the variable responsible for gravitational force, but we have made no such discovery about what is responsible for electromagnetic force. To be electrically neutral is the null disposition to produce no motion in nearby objects, while to be charged ("negatively" or "positively") is to have the power to affect the motion of nearby objects according to certain laws—but if we ask what *grounds* this power, we come up empty handed. Of course, if we are speaking of macroscopic bodies, we can advert to the charges of the constituent particles, so that total negative charge is a preponderance of electrons over protons and positive charge is the converse. But when we ask what it is for an electron to have a negative charge we get nothing except the disposition—the electron is disposed to repel other electrons and to attract protons. What it is *about* the electron that gives rise to that pattern of dispositions is not specified.

Here we reach a difficult metaphysical question: can such a disposition be the end of the ontological line or must it be grounded in some categorical feature of charged objects? This is not the question of whether some *laws* must be taken as primitive;

---

[5] In the old days the two labels were "vitreous electricity" and "resinous electricity" because of their associations with glass and resin, but plainly this gives no insight into the nature of electricity—it merely relates the mysterious force to two types of material associated with it.

I take it that some must. It is the question of whether physical reality can bottom out in brute dispositions. I think it cannot, but I don't want to argue about that familiar issue here.[6] I merely want to note the asymmetry with gravity and mass and to observe that the usual talk of electrical charge is really nothing more than a summary of certain law-governed dispositions—though it certainly gives the impression that it is digging deeper. Textbooks of physics speak as if it is *because* of charge, positive or negative, that interacting bodies move as they do—suggesting that they have a conception of charge that is independent of such motions.

But to have a charge is understood simply as having the ability to induce certain motions in nearby objects—nothing further has been descriptively specified. Explaining electromagnetic motions by citing charges is basically a *virtus dormitiva* explanation, i.e., no explanation at all. At best it tells us that the motions are caused by the force we call "electricity" (i.e., *amber-related*) and not by the force we call "gravity." If we want to know what the *cause* of electromagnetic motions is, we have not yet been enlightened. We merely have a conceptual apparatus to sum up the specific kinds of accelerations observably involved. Electromagnetism, like gravity, is a force that acts at a distance according to an inverse square law, and thus has all the occult qualities associated with Newtonian gravity: but, unlike gravity, we seem not to have any independent grip on its causal powers—we don't know what intrinsic feature of the particle its disposition to attract and repel results from. It simply has a power to cause certain motions, where this power differs from that of gravity—so that the laws are different in the two cases. Gravity is extremely weak for particles, and so produces no appreciable attraction between them; but electromagnetic force is, comparatively, extremely strong between particles—which is fortunate or else atoms and solid matter couldn't exist. Customary expositions of electromagnetic theory, however, tend to disguise the ontological and epistemological status of charge, so that we seem to be dealing in identifiable realities that underlie electromagnetic interactions. If I am right, however, the concepts of electromagnetic theory, as it now exists, are inherently dispositional, merely codifying the distinctive motions we observe. Thus we don't know what electricity *is*—only what it *does*. We have no informative description of its intrinsic nature.[7]

Yet it surely must *be* something, or else nature is just a web of unsupported brute dispositions. It would be different if charge were proportionate to mass or spin or even shape; then we could cite these attributes as the ground of the charge-like dispositions. But charge depends upon no other known attribute of particles: it is fundamental.

---

[6] For a recent discussion, see Peter Unger, *All the Power in the World* (Oxford University Press: Oxford, 2006). Unger and I agree on the mysterious character of the physical world.

[7] So charge takes its place along with matter and motion as an unknown something. We have only an operational definition of charge, in terms of its measurable effects. Charge is just whatever it is that has such and such effects, i.e., certain measurable motions of bodies. Our knowledge of charge is merely structural—dispositional, functional, mathematical. Yet there is (must be) a concrete reality beneath the skeletal sketch: "electric charge" is the name of a (nondispositional) natural kind whose real essence escapes us.

The problem is that it is fundamental without being categorical: that is, the term "charge" is either simply shorthand for the observed motions of (so-called) electromagnetic interactions, in which case it adds nothing explanatory; or else it refers to a "we-know-not-what" that genuinely would explain those motions (if only we knew it)—in which case there is a physical reality out there whose nature we don't grasp. In other words, it's either ungrounded dispositions or profound ignorance. I favor the latter view, as maintained in the rest of this book, in which case electromagnetic theory is a partial and incomplete theory. Put bluntly: the concept of electric charge, as we have it, is just hand waving, however serious and developed the associated laws are. We don't really know what the cement of the universe is—we have no positive descriptive conception of it (as Hume maintained for causation). We have what I have elsewhere called a functional theory of electricity, not a constitutive theory.[8]

There is a further puzzle about electricity that does not afflict gravity: the electromagnetic force is both attractive and repulsive. I don't think it is appreciated how strange and paradoxical this is. Take a single electron suspended in space: if another electron approaches, it vigorously pushes that electron away (allegedly "because" both have negative charge), whereas if a proton comes along it pulls that proton towards it. So the electron is exerting two sorts of influence in virtue of its single nature. Oddly, if you are an approaching electron, you don't experience the attractive force that emanates from the first electron, only the repulsive force—while if you happen to be a proton, you are oblivious to the repulsive force that pervades the space in which you experience a force of attraction. How come every particle doesn't experience *both* forces if both forces are there? It's as if the electron shuts off the attractive force when it senses another electron at a certain point in the vicinity, and shuts off the repulsive force when a proton enters its sphere of influence. So it must "know" what type of particle is in its neighborhood in order to act appropriately—pulling some in while pushing others away (like a nightclub doorman). Gravity treats every body with equal respect, attracting everything with mass, making no distinction between charged masses with different signs.

But the electromagnetic force is strangely invidious. Take an arbitrary point *P* in an electron's electromagnetic field with no particle at *P*, just empty space: what

---

[8] It used to be held that electricity is a kind of fluid, because it can "flow" along conductors. I suspect that part of the attraction of this picture is that it offers some kind of concrete handle on the mysterious stuff called "electricity": we know what fluids are, because we experience them every day, so if electricity is a fluid we have some model of its mode of being. But, aside from other objections, the fluid model gives us no help when it comes to individual particles: charged particles may flow along wires like water droplets along pipes, but what is it for a single electron to have charge? We may know how charge passes from one object to another—by "flowing"—but we still don't know what charge *is*. We know that charge can produce attraction and repulsion, and this we can try to grasp by means of analogy with the human actions of pulling and pushing, but we still don't know what *grounds* these powers. We are reduced to mental images of a certain "glow" surrounding a particle, a ghostly spark perhaps, a strange flickering incandescence. Or we just think of the electric shock we would get if we touched the particle. But none of this adds up to insight into the objective nature of charge: it is just foreign material flowing into the gap created by our ignorance.

force is latent at *P*? Is it attractive or repulsive? We can't say unless we specify which type of particle might occupy *P*: attractive if it's a proton, repulsive if it's an electron. Independently of this, we can only say *P* has a disposition to generate both sorts of movement—it pushes and pulls at the same time. Or should we say it is simply indeterminate what the direction of the force is at *P* until a specific type of particle occupies *P*? But then how can an existing force field be indeterminate in the direction of its action? The field will propel you away if you are one type of particle and suck you in if you are another type, regardless of mass, spin, shape, etc. Really, only *pairs* of particles can be said to be instances of electromagnetic force: an individual particle, considered in itself, seems completely undecided whether it is attractive or repulsive. Since likeness of charge means no more than having a disposition to repel each other, and unlikeness having a disposition to attract each other, we have no *explanation* here of why an electromagnetic field acts as it does—it simply, as a matter of brute fact, attracts and repels simultaneously (or else is indeterminate). This is a very funny kind of force, like Dr. Doolittle's Pushmipullyu, and quite unlike the other forces of nature, which are uniformly attractive. One wants to ask what it is *about* electricity that gives rise to its peculiar duality. What is an electric charge *such that* it produces both attraction and repulsion? But we don't have any real positive conception of the constitutive nature of charge, so the puzzle remains.[9]

In an account of electromagnetic interaction there are customarily four levels of description employed: there is the type of acting particle, say electron or proton; there is its charge or degree of charge, as measured by some conventional system such as voltage; there is the field that surrounds the charged particle; and there is the acted-upon particle—with its own charge and field. The charge is typically distinguished from the field, so that the picture is that the charge *causes* the field, which in turn causes the motion in the particle acted upon. Young and Freedman write: "We first envision that body *A*, as a result of the charge that it carries, somehow *modifies the properties of the space around it.* Then body *B*, as a result of the charge that *it* carries, senses how space has been modified at its position" (806). That is, a field in space is created by the presence of a particle endowed with charge, and that field transmits the effects of charge to other particles. But it is not clear that we are entitled to the distinction between charge and field, or how that distinction is to be drawn: for how is having the charge to be physically distinguished from having the field? The matter of a body is distinct from the gravitational field surrounding it,

---

[9] The cement is also explosive, as it were—a source of dispersion as well as cohesion. What is electric charge? An I-know-not-what that pushes and pulls, that has contrary effects on affected bodies. Why don't other forces have such a dual nature? Gravity always pulls; it doesn't have a repulsive form. What would we think if the sun attracted the earth but repelled Saturn? What if gravity sometimes made objects fly up from the earth and sometimes pulled them down? That would be very perplexing, would it not? But charge does this all the time—now pulling, now pushing. Why can't it make up its mind?

obviously, but how is the charge to be separated from its associated field (I don't mean the particle itself, but the charge it carries)? Isn't having the charge simply *identical* to the field? What is to stop us from saying this? Does it make sense to say that a mass has both a "gravitational charge" *and* a gravitational field?

The very fact that this is an ontological option, to be determined by considerations of parsimony and the like, shows how little grasp we have of the nature of charge. Perhaps we should say that *for all we know* charge and field are identical. But if they are, charges cannot *cause* fields. I myself would prefer the idea that charge is an intrinsic and local feature of particles whose nature we don't know, so that charges can cause fields; but our grip on charge now is so slender that it is hard to be dogmatic on the point. In any case, holding to an identity theory hardly resolves the puzzles of charge, since electromagnetic fields are just as perplexing. Are they to be conceived purely dispositionally, specified by categorically ungrounded counterfactuals, or are they physical realities with an intrinsic nature?[10]

I lean to the latter view, but then we have to admit we don't know what their nature is. The whole issue bristles with difficulties and obscurities, showing how little we really understand the ontological standing of the things we call "charge" and "field." In fact, I think that the physicists simply observed certain kinds of interactions, decided to call them "electrical," postulated an attribute they labeled "charge," brought in associated "fields," and then proceeded to work out the laws that govern those interactions; what exactly all these terms denote was left completely unsettled. When the laws proved mathematically expressible and strongly predictive, the theory was deemed acceptable—despite the epistemological gaps it contains. Their work as scientists was done. Much of the theory is therefore conventional and stipulative, despite having the surface appearance of empirical law and genuine physical reality. Of course, there is nothing wrong with it as science, but a determined metaphysician (or reflective physicist) will sense many unanswered questions (and I would suppose the metaphysician in the practicing physicist to be not entirely dead, despite the operationalist beating he or she may have received[11]).

---

[10] See Marc Lange, *An Introduction to the Philosophy of Physics: Locality, Fields, Energy, and Mass* (Basil Blackwell: Oxford, 2002) for a discussion of realism and fields. He favors a realist view of fields, as opposed to a counterfactually-based view, as do I: the field is really the categorical ground of the associated counterfactuals.

[11] It is not as if operationalism is free of internal difficulties, whatever one might think of it as a general philosophical stance. It boasts a rigor, clarity, and explicitness that are supposed to contrast with airy-fairy realism, but it quickly disintegrates under careful scrutiny. Take length, the standard best case for the operationalist viewpoint: length is held to be what is measured by an appropriate measuring apparatus. But what kind of apparatus is that? A foot rule or a tape measure will work for shorter lengths, but what about the distance from the earth to the sun? Here we have to use such measures as the speed of light. So is the concept of length *ambiguous* between the two types of measurement? Is distance what is measured by applying a rigid rod or what is measured by using the speed of light? Intelligence can be measured in several different ways, so are there as many types of intelligence as there are ways of measuring it? And what exactly is a

Three final points should be made. First, most of us have suffered an electric shock, if only by means of "static." This may make us think we have had a direct experience of what electricity is like, so that we know what it is by acquaintance. But that is clearly a naïve view of the matter: all we have experienced is the effect of electricity on the nervous system, not its intrinsic nature (compare our experience of heat and its real nature). Such experiences do not reveal what electricity is in itself and objectively (as, by contrast, experiences of red *do* produce knowledge of what redness is in itself, or at any rate knowledge of what experiences of red are). Electric charge has a disposition to generate experiences of a certain qualitative character in humans, but these experiences cannot be construed as revealing the intrinsic nature of electricity, so they don't fill the theoretical gap I identified above (any more than experiences of heat can tell us that heat is molecular motion or experiences of light can tell us that light consists of photons in motion).

Second, electrostatics is the study of electric charges at rest, while electrodynamics deals with charges in motion. It is known that only charges in motion produce magnetic fields; static charges are magnetically neutral. Of course, the basic charged entities of nature, such as electrons and protons, are in constant states of motion, so their magnetic properties are constantly active too. But it is worth pondering the oddity of the empirical facts here: how could simply moving a charged particle produce a magnetic field where none existed before? Mere displacement doesn't usually change the causal powers of an object, but in this case a new kind of causal power

---

measurement? What does the operation of measuring consist in? Is it a physical operation, like placing a measuring rod next to the measured object? Then is this physical operation involving the rod also subject to operationalist interpretation—as it might be, measuring the physical fact of measurement? That way a regress lies. So is measuring better understood as a *mental* act—judging, say, that a person is a certain height or calculating the distance of the sun? Then we are explaining the content of physical concepts in terms of mental concepts, with all the mind-dependence that goes with that: for a physical thing to have a certain length *is* for certain mental operations to be performed or be performable. That sounds quite wrong. Or is measurement to be conceived purely mathematically—as a mere mapping of numbers onto objects? The physical objects stand in certain constitutive relations to abstract entities that record their magnitude. But then we are explaining the concrete by the abstract—physical facts in terms of facts about numbers. In that direction a bizarre Pythagorean Platonism about the physical world looms: physical things get reduced to mathematical things. The plain fact is that physical things like length are not *identical* to bodily acts of apparatus application, nor to mental acts of length cognition, nor to the numbers used in scales of measurement. And the same goes for mass, motion, charge, and so on. Measurement, however we choose to construe it, is *extrinsic* to what is measured, and so cannot constitute what is measured. In fact, the entire approach is viciously anthropocentric, and hence idealist in upshot. Measuring is what humans do; nature does not measure itself. The essence of physical reality, as an independent domain, cannot then consist in measurement. The grain of truth in operationalism is simply that if we can't measure something we can't formulate quantitative laws about it, which is what physics aims to do: but this has nothing to do with the constitution of physical reality as it is in itself.

comes into existence, with distinctive laws and principles—simply by moving the particle from one place to another. And this is particularly puzzling when you consider that the motion in question is generally held to be relative, so that whether something is declared at rest is essentially arbitrary: this means that the magnetic field is produced only *relative* to other objects and not to the object under consideration (it is at rest relative to itself and hence magnetically neutral relative to itself). On the other hand, if motion is absolute, mere displacement in empty space can produce a new force, and whether a solitary object in space has a magnetic field depends upon whether it is in motion in relation to absolute space. Perhaps if we knew better what charge was, we would understand how the motion of charged bodies gives rise to a magnetic field; as it is, this has the look of a brute fact, and a counterintuitive one at that. After all, people don't become more intelligent or better athletes just by being set in motion. How can mere motion change the causal powers of an object?

Third, there is the question of the electrically neutral bits of matter. The atomic nucleus is said to consist of positively charged protons that naturally repel each other (hence the need for the "strong force" to keep them from flying apart) and also neutrons that have no electric charge—sit them next to an electron and nothing happens, attractive or repulsive. So nature consists ultimately of particles that have charge, that attract and repel, and those that have no charge, that neither attract nor repel. Yet all three consist of matter and possess mass and other basic physical properties. It seems that the neutrons just primitively lack charge, as the electrons and protons primitively have it. Could the neutrons be exactly like the protons in all physical respects *except* the possession of charge? If so, charge is not supervenient on those other physical properties. That seems consistent with physics as we have it. But again, this seems very odd and cries out for explanation: what *makes* a neutron lack the disposition that defines charge for protons and electrons? And why is charge found so generally in nature and yet not for this particular class of particles? It can hardly be *in order* that atoms can hold together, since no such teleology obtains in atomic nature (or cosmic).

Once more, we are confronted with a brute fact—or at least the appearance of brute fact. One would think that the difference between electrical neutrality and electrical charge, which matters so enormously to the behavior of particles, would reflect something about their intrinsic nature; yet so far as we know the difference is brute—some things have the disposition to engage in electromagnetic interactions and some don't, *period*. Maybe, from a wider perspective, the mysterious thing is neutrality not charge—as if the general rule of nature meets an exception in the case of the neutron (electrons and protons look at neutrons with blank incomprehension, finding their own charged state the natural state of things). In any case, the holy trinity of electron, proton, and neutron form a very mixed and puzzling bunch, not much less mysterious than other trinities we might think of. Once again, successful science goes along with natural mystery. The theory of electricity and magnetism is one of the most impressive and successful areas of human knowledge, but still it conceals vast chasms of ignorance.[12]

---

[12] That is why I subtitled this paper "A Case Study": because it illustrates the general point that an advanced branch of physics, rigorously mathematical and rich in applications, can exist in the absence of any real grasp of its own subject matter. In the case of electromagnetic theory, moreover, there is a kind of illusion of deep understanding, because of all the talk of positive and negative charge, of conductors and fields, of dipoles and flux: this gives the impression to the novice that the physicist is in full possession of the whole truth about electricity and magnetism, right down to their inner nature. But in fact our knowledge of electricity is glancing and sketchy—dispositional-cum-mathematical-cum-structural. It is a kind of castle in the air: a magnificent structure floating on inscrutable foundations.

8 Two Types of Science

A natural taxonomy of the empirical sciences would break them down into three basic groups: the physical sciences (physics, astronomy, chemistry, geology, metallurgy), the biological sciences (zoology, botany, genetics, paleontology, molecular biology, physiology), and the psychological sciences (psychology, sociology, anthropology, maybe economics). Why does such a threefold scheme seem natural? Because each science employs a distinctive basic concept that defines its domain—what it attempts to describe, explain, formulate laws of, understand. In summary form, we can specify the operative concepts as the following three: *force, purpose,* and *mind*. Thus physics deals with the forces of gravity, electromagnetism, and whatever other forces control the motions of matter; specifically, it gives the laws, mathematically stated, that govern these forces. Biology organizes itself around the notion of purpose or function—teleological notions. Organisms aim at their own survival and reproductive success, while bodily organs have a function in promoting the life of the organism. Psychology, trivially, deals with the mind in its various manifestations—with thinking, emotion, perception, memory, etc. It is a controversial question whether these various departments of psychology can be united under a single heading—such as consciousness or intentionality or information processing—but clearly the concept of the mental is what defines the psychological sciences.

Setting aside, for the moment, questions about how our three broad concepts—*force, purpose,* and *mind*—might be adequately defined, we can indicate the concerns of the various sciences by way of example. Thus physics is interested in, for instance, electrical and magnetic force; biology will inquire into the function (and functioning) of, say, the heart; and psychology will deal with visual perception, for example. These more specific concepts—*electromagnetic force, the heart,* and *vision*—illustrate the region of empirical reality the science in question is concerned to understand. They are the *kinds* of concept that the relevant science is organized around.

I am interested in the nature of these concepts as cognitive entities, and what their cognitive nature tells us about the sciences that employ them. More exactly, I am interested in what kind of *knowledge* is possessed when these concepts are deployed. And my thesis is to be that two very different kinds of knowledge are possessed—one kind in the physical sciences, another in the biological and psychological sciences. There is no accepted vocabulary for the distinction I have in mind, so I shall introduce the following technical terms: "*remote* knowledge" and "*intimate* knowledge." My thesis, then, is that in physics we have remote knowledge, while in biology and psychology we have intimate knowledge. The significance of these labels should become apparent as we proceed. The general upshot is that physics is a cognitively different kind of science from the psychological and biological sciences: the characteristic concepts of the two types of science are fundamentally different.[1]

In *The Nature of the Physical World* (1928)[2] Eddington speaks of "the essential unknownness of the fundamental entities of physics," comparing the statements of that science to the lines of Lewis Carroll's "Jabberwocky." "*Something unknown is doing we don't know what*," he writes—rather as "The slithy toves/Did gyre and gimble in the wabe" (291). He distinguishes three things: "(a) a mental image, which is in our minds and not in the external world; (b) some kind of counterpart in the external world, which is of inscrutable nature; (c) a set of pointer readings, which exact science can study and connect with other pointer readings" (254). He then speaks of "the limitation of physics to pointer readings," with the inscrutable physical counterpart consigned to obscurity. Thus: "Whenever we state the properties of a body in terms of physical quantities we are imparting knowledge as to the response of various metrical indicators to its presence, *and nothing more*" (his italics, 257). Accordingly, "we realize that science has nothing to say as to the intrinsic nature of the atom. The physical atom is, like everything else in physics, a schedule of pointer readings. The schedule is, we agree, attached to some unknown background" (259). When it comes to discovering the intrinsic properties of nature, "we [physicists] run round and round like a kitten chasing its tail and never reach the world-stuff at all" (280). He concludes: "It would not be a bad reminder of the essential unknownness of the fundamental entities of physics to translate it into 'Jabberwocky'; provided all

---

[1] We could also say that the meaning of the terms is different: how they are introduced, the conditions for understanding them, the shape of their sense.

[2] The book is based on Sir Arthur Eddington's 1927 Gifford lectures, published originally by Cambridge University Press, in 1928. It provides a superbly clear and lively account of the developments in physics of the early twentieth century, showing much philosophical sensitivity.

numbers—all metrical attributes—are unchanged, it does not suffer in the least. Out of the numbers proceeds that harmony of natural law which it is the aim of science to disclose. We can grasp the tune but not the player. Trinculo might have been referring to modern physics in the words 'this is the tune of our catch, played by the picture of Nobody'" (291–2).

In *The Analysis of Matter* (1927)[3] Russell takes much the same line. "It is not necessary to the physicist to speculate as to the concrete character of the processes with which he deals" (122), he writes. This is fortunate because "we know nothing of the intrinsic quality of the physical world, and therefore do not know whether it is, or is not, very different from that of percepts" (264). On the contrary: "The only legitimate attitude about the physical world seems to be one of complete agnosticism as regards all but its mathematical properties" (270–71). Physics is concerned only with what Russell calls "structure," i.e., purely formal aspects of concrete things, and hence is entirely "abstract." Physics studies only the "causal skeleton of the world" (391), he says, leaving its concrete flesh unrevealed. "We know the laws of the physical world, in so far as these are mathematical, pretty well, but we know nothing else about it" (264). That is why we don't know enough about physical reality to rule out the view that its essence is experiential (which is also Eddington's position). It is as if in describing an array of colors we were limited to their abstract relations of similarity and difference, without being able to grasp the identity of the individual colors. For Russell, physics is epistemologically removed from the realities to which it refers: it does not inform us, by acquaintance or description, what the physical world is like in itself; instead, it provides a mathematical interpretation of various lawlike relations among physical phenomena.

Although Eddington and Russell appear to have converged on the same view independently, both may be presumed to have known about Poincare's early statement of the view in *Science and Hypothesis* (published in English in 1905)[4]—though neither acknowledges his influence. Speaking of the equations of physics, he writes:

these equations express relations, and if the equations remain true [under theory change] it is because these relations preserve their reality. They teach us, now as then,

---

[3] Dover Publications: New York, 1954. Widely neglected by mainstream analytical philosophers for decades, Russell's book is now receiving its due. See also Russell's *An Outline of Philosophy* (George Allen and Unwin: London, 1927), in which he says: "Physics is mathematical, not because we know so much about the physical world, but because we know so little: it is only its mathematical properties we can discover. For the rest, our knowledge is negative" (163–64).

[4] I quote here from the edition of *Science and Hypothesis* published by General Books, 2009. The more standard edition is by Dover Publications, 1952. There are some interesting differences in the translations of these two editions: for instance, my edition uses the phrase "foundation of things" while the standard edition has "causes of things." Also my edition has an appendix called "The Principles of Mathematical Physics," from which my third quotation is taken, while the standard edition does not have this appendix. The book is rightly celebrated for its pithy insights and scientific expertise.

that there is such and such a relation between some thing and some other thing; only this something we formerly called *motion*; we now call it *electric current*. But these appellations were only images substituted for the real objects which nature will eternally hide from us. The true relations between these real objects are the only reality we can attain to, and the only condition is that the same relations exist between these objects as between the images by which we are forced to replace them. (100)

Later, in dealing with the problem of multiple explanations of the same experimental data, he remarks: "A day will come perhaps when physicists will not interest themselves in these questions [such as whether mechanism can be made to work], inaccessible to positive methods, and will abandon them to the metaphysicians. This day has not yet arrived; man does not resign himself so easily to be forever ignorant of the foundation of things" (130).

In a striking passage in which he compares the universe to a machine whose inner workings are not open to us, he says: "This [the universe] is also a machine, much more complicated than all those of industry, and of which almost all the parts are profoundly hidden from us; but in observing the movement of those that we can see, we are able, by the aid of this principle, to draw conclusions which remain true whatever may be the details of the invisible mechanism which animates them" (145–6). Here we see the germ of what I have called the functionalist view of physics.[5] Poincaré's position might be called "relationalism"—the view that physics only provides knowledge of relations between things, while leaving the nature of the relata obscure and hidden. He takes our profound ignorance of nature for granted, and then tries to explain how physical knowledge is still possible—because we can ascertain the relations that obtain between things and put them into mathematical form. But we should be under no illusions about what physics does *not* tell us—namely, the "foundation of things."

This type of view is now known as the "structuralist" view of physics, and it has been championed by many others, early and late.[6] I won't try to defend it further

---

[5] Roughly, this is the idea that the intrinsic nature of physical entities is left unspecified, while their interrelations are mapped: the entity is described by a definite description of the form "the entity with such and such causes and such and such effects." The formal version of this analysis is the Ramsey sentence formulation of the content of a theory, replacing constants by existentially quantified variables. Ramsey originally suggested his analysis for physical theories, not psychological theories—and with good reason, since it is precisely wrong for psychological theories, as I shall be arguing in this paper. The Ramsey sentence formulation of physics fits the Poincare-Eddington-Russell position perfectly.

[6] As I observed in the Introduction, note 2, Hertz preceded Poincare in stating the basic idea. But the position has clear roots in the writings of Locke, Hume, and Kant: all three were dedicated to charting the extent of human knowledge, and all detected large areas of ignorance. I particularly like Galen Strawson's presentation of the position: see his "Real Materialism" and "Realistic Monism" in *Real Materialism and Other Essays* (Oxford University Press: Oxford, 2008), with references to Michael Lockwood and others (it fits Strawson's persuasive account of Hume as an agnostic realist). John Worrall is generally credited with introducing the idea of "structural realism" into current philosophy of science, acknowledging the influence of Poincare: see his "Structural Realism: the Best of Both Worlds," *Dialectica* 43, 99–104, 1989. Perhaps my favorite short statement of the position

here: my aim is to adopt it and show its significance. But I will say something by way of interpretation. The view is comprehensive: the kind of unknownness of which Eddington speaks afflicts all of basic physics—electromagnetism, gravitation, the atom, matter itself, space and time, even motion. All these terms are really just labels for things whose objective nature is not directly revealed to us and of which we have no positive conception. What physics gives us is a mathematical language for talking about these elusive "slithy toves," and with it a set of laws: but what we really have in the relevant theories are measures that are connected to other measures—an elaborate system of "pointer readings," as Eddington puts it. What he calls the "background" to these measures is hidden; only the effects of the labeled quantities on our measuring instruments are directly known to us—and even these instruments, as physical entities, have a hidden objective nature. It is the mathematics that is given to us—the numbers we assign to the measuring devices—and this is entirely abstract.[7] Physics consists of equations defined over unknown entities. Thus we don't know what an electron *is*: we know only what it *does* according to the laws of particle physics (and, of course, these are very odd, in view of the quantum theory).

Eddington calls our knowledge in physics "symbolic knowledge," because it relates only to symbols of things whose nature we do not grasp. I call it "remote knowledge," because it concerns the remote effects of things on our senses (as they process our measuring instruments). It might also be called "functional knowledge" (in the sense associated with functionalism) in that it leaves unspecified what the basic entities are intrinsically, detailing only their functional relations to other unspecified entities. It is a mathematical formulation of the behavior of things that elude our grasp as to their intrinsic character. To be a little more concrete: matter consists of charged particles, according to physics, but we have nothing more to say about charge than is specified in a set of equations concerning motion—where this motion (at the atomic level) is quite other than what we experience when (say) a bird flies by. The theoretical terms contain no more descriptive content than the equations in which they figure, being merely names for physical realities that we fail to grasp. The kind of

---

comes from Lewis Carroll Epstein's lucid *Relativity Visualized* (Insight Press: San Francisco, 1997): discussing the nature of light, he observes that at the time Einstein formulated Special Relativity it was unknown what light is—and then he suavely inserts the parenthesis: "(In fact, it is unknown what anything really is)" 23. Still, he goes on to give a nice exposition of Relativity Theory.

[7] Also, of course, we know the mental effects of the unknown physical quantities. Perhaps this knowledge is the chief source of the illusion that we know the nature of their causes: we have a solid grasp of the effects of matter on our consciousness, so we mistakenly infer that we know what matter is itself. The nonsequitur is apparent, if not quite numbing, but also tempting. It can also be said that we know the nature of the numbers assigned as measures of physical quantities: but again, this is not to say we know the nature of the things to which those numbers are assigned. As Russell would put it, we have knowledge by acquaintance of the surroundings or correlates of external physical entities, in the form of mental effects and mathematical measures, but our knowledge of the physical entities themselves is merely under structural descriptions *linking* them to what we are acquainted with.

knowledge we have of the electron, say, is like the knowledge we have of a man when we have seen only his footprints—merely knowledge of effects.

In general, the terms of physics are definable as "the cause of F," where "F" cashes out ultimately as pointer readings or some such. This is why Eddington feels able to claim that physics leaves it an open question whether matter really consists of "mind stuff": it is simply too *unspecific* to rule out such an hypothesis. The actual cause of our pointer readings might be anything, so long as it "realizes" the right mathematical laws. It is not merely that the entities of physics are unobservable and known to exist only by inference: their very *nature* is unknown to us—what kind of thing they intrinsically *are*. Thus a great ignorance lies at the heart of physics—though it is an ignorance that does nothing to impede the predictive power of the science. We know with mathematical precision what our slithy toves will do in the future (subject to qualifications about determinism), though we are in the dark about what their slithiness consists in and what their tovehood amounts to.

It should be noted that this agnostic structuralist position has nothing essentially to do with a rejection of naïve realism about human perception (though Russell sometimes writes as if there is such a connection). The point is not that we don't or can't perceive material things directly and hence can only know their properties inferentially; there is no connection with sense-datum theories of perception and the like. I myself don't question the commonsense view that we can see tables and chairs and so on—the very things that physics deals with. We can also know many truths about the physical objects thus perceived—shape, color, location, etc. The question is whether we can know their intrinsic nature as physical objects—as objects of physics.

Suppose (wrongly) that we don't really ever see material things but only what Russell calls percepts: it simply doesn't follow that we can't know what the intrinsic properties of matter are. We can't see other people's mental states directly either, but we *do* know their intrinsic character—what they *are*. The question is not whether we can perceive physical objects "directly"; it is an issue about our *conception* of what we refer to, either perceptually or inferentially. If we lived in a very different physical world, in which matter worked more like the classic mechanist picture, or in which Eddington-Russell panpsychism were true, then we might well find such a world intelligible, even though in that world our perceptual system for some reason rendered matter unperceivable "directly." Knowledge by inference is not condemned to be merely structural: it depends on what the knowledge is *of*. And even if we accept (rightly) that *our* matter can be directly perceived—in the sense that parcels of it can be seen and touched—it doesn't follow that our understanding of its nature should be geared to reveal its inner being. There is no entailment from being able to directly perceive congeries of electric charges (this being what matter basically is) to understanding the nature of electricity. That it should be correct to describe us as in perceptual contact with material things, instead of sense data, gives no ground for supposing that our physics can provide insight into what material things ultimately are. And, of course, this is especially obvious if the appearance of material things to us involves distortions and errors as to what matter is really like—as with the amount of empty space inside a typical "solid" object (a theme of Eddington's). We might really be able to see and touch lumps of matter but have no idea at all about what matter is intrinsically like as revealed by science.[8] The structuralist point concerns the interpretation of the theoretical terms of physics, not whether our senses take material things as direct objects.

There is a great contrast between physics and psychology in respect of epistemological accessibility. Psychology is notoriously backward as a science compared to physics: there is none of that mathematical precision and predictive power, as well as fine-grained analysis of structure. The physicist, we have been schooled to think, knows so much more about the physical world than the psychologist knows about the mental world—the laws, in particular. In a sense, that is quite true—but it is also importantly false.

The reason is that we *do* know the intrinsic nature of our mental states—our conscious thoughts, feelings, sensations, and so on. This is an old and hoary point: conscious experience is indubitable and the nature of particular conscious states is given to the subject.[9] I just do know what pain is, and seeing red, and feeling angry, and thinking it's about to rain. I have *intimate* knowledge of such things, in the sense that I know what they are like intrinsically. I am not confined to *inferring* their presence from perception of their effects and then blankly dubbing them with some name lifted from Latin or Greek (as with "electricity"). They are not remote from my cognition, hidden variables, obscurely lurking in the background, but are transparently present to it. What their causes and effects may be, or their neural correlates, is a matter of speculation: but as to what they are essentially, that I am well aware of (as I am aware of geometrical forms, for example). Words for mental states are *not* like "slithy toves." We know the *identity* of our experiences—what they are in themselves. There is, in this sense, nothing theoretical about our mental

---

[8] Presumably, if naïve realism is true of human perception, it is true of animal perception: if we don't see sense-data, then apes and dogs don't see sense-data either. But of course it doesn't follow that animals know the nature of matter, motion, electricity, etc. We (and they) know how matter *appears*, by really seeing physical objects, but that is not the same as knowing what matter intrinsically is. Nor does a form of intelligence necessarily have to be in perceptual contact with objects in order to grasp their inner essence: God presumably grasps it, but I doubt that he does much seeing and touching of things (he "apprehends").

[9] Strawson gives forceful expression to the point in "Real Materialism," cited in note 6. It is a point that seemingly needs to be rediscovered every century or so. Really the point tells us something about the powers of introspection, not so much the nature of experience per se—because experience is not *in itself* any more transparent than anything else. God is as intimately acquainted with electric charge as he is with conscious experience—it is as if he can *introspect* electric charge. The asymmetry between charge and experience for humans arises from the way introspection applies to the latter but not the former, I would say. Still, experience is such that it *can* be known by introspection, while charge is presumably not. (There are some difficult questions in this vicinity.)

terms.[10] We don't know mental states via pointer readings and fancy apparatus, but simply by having them. The mind is not an "I-know-not-what." The concept of belief, for example, is thus quite unlike the concept of electric charge: we understand the two concepts quite differently. The ancient Greeks may have discovered electricity and given it a name, but it would be decidedly odd to say that they discovered belief and gave it a name. The former is postulational knowledge; the latter is immediate knowledge.

I take it this point is familiar enough (though not always accepted). We can put it by saying that our knowledge of the mental is not purely "structural," not completely "abstract," not inherently "relational," not limited to the "functional." None of the formulations of the character of physical knowledge offered by Poincare, Eddington, and Russell carry over to psychological knowledge. The lesson I want to draw from the point is less familiar—namely, that psychology as a science rests on conceptual foundations that differ in kind from those of physics: psychology is about something whose intrinsic nature we do know, while physics is about things whose intrinsic nature we don't know (to put it crudely). In psychology we know *what* we are talking about, but in physics we don't—despite the brilliance of what we *say* about those things whose intrinsic nature we don't know. We have intimate knowledge in the one case and remote knowledge in the other: or, in an earlier terminology, we have "adequate" ideas of the mental and "inadequate" ideas of the physical: or, in yet other terminology, we have an "intelligible conception" of the mental and an "unintelligible conception" of the physical (though a perfectly adequate predictive theory of it).[11]

When we apprehend the laws of physics we have a gap in our understanding corresponding to the nature of the entities governed by those laws, but when we apprehend the laws of psychology we know perfectly well what the nature of our subject matter is. For instance, we know the laws of electric charge, but we don't know what electric charge is, except dispositionally. By contrast, when we know the laws that govern belief and desire we are quite clear about what belief and desire are.

---

[10] Thus the content of psychological statements cannot be captured by the Ramsey sentence paraphrase. People failed to grasp this point in the heyday of functionalism because they insisted on taking a third-person perspective on the mind, viewing mental states as theoretical posits introduced to explain observable behavior. My short reply to this: Rubbish! We did not *invent* our ordinary psychological concepts as part of a lengthy endeavor to bring order to what we observe of other people's behavior, with many a false start, paradigm shift, and theoretical controversy. This is to ignore completely the way we are epistemically related to the contents of our own minds. It is really quite amazing that it is (still!) necessary to make this crushingly obvious point—such is the power of ideology to override simple good sense.

[11] These terms are from Locke and Hume, respectively. They were much concerned with the kind of contrast I am harping on: which things we really grasp as they objectively are and which things we don't grasp but merely refer to. The crucial point is that we can refer to things, or think *about* them, and yet have no proper descriptive conception of them: they exist, we refer to them, but the nature of their existence is closed to us. See Strawson's discussion of Hume in "David Hume: Objects and Powers" and "Epistemology, Semantics, Ontology, and David Hume," in the volume cited in note 6.

I am not here saying that the epistemology of our knowledge of the mental is completely clear—obviously, this is a very problematic area. My point is not that self-knowledge is easy to understand or free of philosophical perplexity. Nor am I saying that nothing about the mind is mysterious: there is much ignorance here, both superficial and deep. My point is the very modest one that our ordinary mental concepts do not leave us in the dark about the intrinsic character of mental states. Our mental terms do not merely *point* to an unknown reality; they *express* a known reality. In this respect, the terms "matter" and "consciousness" are very different—with all the mystery attaching to the former.[12]

What should we say about the study of language? Does it give remote knowledge or intimate knowledge? Linguistics deals with syntax and semantics, as well as pragmatics, attempting to derive general principles. Suppose we are interested in nouns: we know quite well what these are, I suggest, because we have direct experience of them, in the production and reception of speech.[13] They are not merely inferred entities whose nature escapes us—we don't know them merely by their effects. The "parts of speech" are not like the parts of matter—remote postulates. Meaning (and grammar) is not like matter: a mere "background" to what we use to measure it, consisting of we-know-not-what. Our knowledge of language is not exclusively mathematical and abstract, a mere schedule of equations, devoid of specific concrete content—remotely connected to language itself. The meaning of a word is not like the electric charge of a particle, epistemologically speaking. Meaning is like mind (unsurprisingly) and unlike matter and charge (and even motion). The atoms of language (say, proper names and simple predicates) are nothing like the atoms of matter, epistemologically speaking: the former are presented to us in their essence—raw, as it were—but particles like electrons and protons are conceptualized merely in terms of structural properties—as centers of force, basically, where force is captured by abstract measurements.

In psychology and linguistics we have concrete positive conceptions of what we are talking about; in physics we have mere labels for radical unknowns. As Eddington remarks: "Our bodies are more mysterious than our minds—at least they would be,

---

[12] I know how matter appears to me in sensory perception, without knowing what matter is; but it would be wrong to say that I know how consciousness appears to me in introspection, while not knowing what it is. The reason is that the appearance of consciousness is (at least part of) what consciousness is, but the appearance of matter is not (even part of) what matter is. You can't know the appearance of consciousness without knowing what consciousness is, but you can know the appearance of matter without knowing what matter is. The gap between appearance and reality for matter does not have a counterpart in a gap between the appearance of consciousness and consciousness.

[13] We may also be supposed to have a grasp of what a noun is by means of our innate language endowment: our inborn linguistic competence. Language is a human trait, written into our nature, internal to us—not an external reality whose nature we must laboriously discover, like the particles of physics or the solar system. Thus we have a privileged epistemic relation to the categories of language: grammar is transparent to us in a way that the atom (say) is not. No one supposes that physics is known innately (theoretical physics, that is).

only that we can set the mystery on one side by the device of the cyclic scheme of physics, which enables us to study their phenomenal behavior without ever coming to grips with the underlying mystery" (277). This is emphatically not true of psychology and linguistics (or sociology, economics etc.), as Eddington and Russell are both quick to insist. We know experiences in their concrete reality, not as the conjectured origin of measurements; and much the same can be said of meanings and the categories of language. This is not to say that the *theory* of meaning (or grammar) is easy or devoid of perplexity—far from it; it is only to say that the *kinds* with which such theory deals are familiar to us—negation, conjunction, quantification, reference, predication, truth, assertion, command, and so on. Our knowledge of these things is not conjectural and remote, mere blank labeling, mediated only by effects and symptoms. We know, say, what denoting *is*, as we know what pain is, even though we may not have a good theory of it; by contrast, we don't know what electrical attraction is—though our theory of it is sophisticated. The epistemology of linguistics is thus quite different from the epistemology of physics.[14] Our grasp of the word "meaning" is not like our grasp of the word "matter": the former designates a reality we have encountered in its intrinsic being, but the latter is an empty label for something we-know-not-what that transcends our apprehension—merely that which underlies the appearances. In Kantian terminology, meaning belongs to the phenomenal world, while matter (real mind-independent objective matter) belongs to the noumenal world. Meaning, after all, depends on us for its existence, but matter is the mind-independent substance par excellence—that which precedes and conditions human existence. Matter is not a product of mind, but the primordial foundational stuff of the universe. Hence the epistemological asymmetry I have noted.[15]

Biology presents an interesting case: on which side of the great epistemological divide does it fall? Well, what are its basic organizing concepts? They are *function, purpose, design, survival, selection, health, flourishing, fitness, life*. What is immediately noteworthy about these concepts is their affinity with the psychological.

---

[14] The linguist articulates categories that are implicit in our ordinary grasp of language—which we know tacitly; but the physicist does not merely articulate what people naturally know of the physical world, making explicit what was implicit. The physicist has to go beyond what is already contained in human knowledge, if only tacitly—which is why radically innovative ideas may need to be introduced. In a sense, linguistics introduces no genuine novelties (in so far as it seeks to articulate pre-existing linguistic competence), while physics inevitably does so, because it goes beyond what is already cognized (same for chemistry, biology, history, and so on). Thus we know language more intimately than we know physical nature, as both theorists and regular people.

[15] I think much the same story can be told about the theoretical categories of sociology, anthropology, and economics—though I will leave that for another occasion. In brief, consider such concepts as *conformity, ritual,* and *recession,* as typical of those three disciplines: these concepts are not like *charge* or *motion* or *matter*, because they have clear descriptive content—we know what they refer to. Asked what conformity, ritual or recession are, we can give a perfectly adequate reply (even if we don't have good predictive theories of them)—we are not confined to a blank I-know-not-what reply. These are not mere labels for a natural kind whose essence escapes us. Our knowledge of their reference is not merely structural/functional/mathematical.

Surely we understand these terms, initially at least, by way of such notions as *intention, aim, well-being, choice, pleasure, plan, self*, and so on. In fact, for organisms that have rich forms of consciousness, the biological terms map precisely onto psychological terms: I *intend* to survive, am *concerned* about my health, *aim* to keep fit, *choose* a mate, *want* to have children, *feel* happy, etc. In general, I have *purposes* and these relate to my well-being and ultimately my survival (i.e., the persistence of my self or consciousness). The biological teleology is correlated with psychological teleology: the ends of my organism are also *my* ends. Thus this notion of purpose, applicable to whole organisms, is securely psychological, and therefore inherits the transparency of psychological concepts. Accordingly, the part of biology that employs this concept exhibits intimate knowledge in my technical sense: we know what the property is that we are referring to—the property of having a purpose (a desire, an intention, a plan). I think the typical way that biological theories are understood trades on this assimilation to the psychological, and hence there is no feeling of mystery at the ground level—as if we were merely gesturing at something whose intrinsic nature eludes us, as with physics. If biological purpose is simply purpose in the ordinary psychological sense, then biology belongs on the side of psychology. For instance, parental care behavior in animals would be interpreted as resulting from the aims and intentions of the animals in question.

However, there is clearly more to biology than that—if only because of plants. That is, organisms that lack a psychology also fall under the laws of biology: they too exhibit function, even purpose, as well as survival, flourishing, reproductive fitness, and so on. The teleological extends beyond the literally psychological. Now it may be that such notions are commonly used metaphorically, self-consciously so, by extension from literal paradigms: but biologists surely believe that they can be defined in such a way as to remove the element of metaphor. At the very least, we need to be able to apply the notion of function to *parts* of organisms, such as internal organs, and they certainly don't have purposes in the psychological sense. And you can readily see how this is supposed to go: a part of an organism is said to have a function if its presence is explained by its contribution to past fitness, or some such condition.[16] Survival of organisms is not exclusively a matter of a conscious self persisting through time (i.e., the opposite of death as we understand it in our own human case) but can be understood in nonpsychological terms—as, say, continued physiological integrity or some such thing. Plants survive and perish, after all.

Let us grant, then, that the characteristic concepts of biology can be cleansed of all psychological connotations, so that they can be used to cover both conscious organisms and the other kind, as well as parts of organisms. *Then* does biology become like physics or like psychology with respect to its epistemological status? I think it

---

[16] For a survey, see Colin Allen's entry on teleology in biology in the *Stanford Encyclopedia of Philosophy*, 2003. Larry Wright and Ruth Millikan, among others, have both made significant contributions.

stays with psychology, because the key organizing concepts, now freed of psychological associations, can be understood by means of the explicit definitions that have become attached to them by stipulation. The notion of function is now defined by reference to past evolutionary success or some such, and survival is now detached from all ideas of psychological persistence and replaced by the idea of physiological integrity. It is not as if by using the terms in the defined way we are simply pointing blindly at things whose nature we can't specify, as with electric charge and the like: we do know what we mean by the terms, because we know what function or survival *are* under the new dispensation. Knowing that the heart has the function of pumping the blood is *not* like knowing that the electron has negative charge.[17]

But notice that even if a residue of the old psychological meaning clings to these terms, we still get the result I am contending for, since biology will now clearly fall in with psychology and not with physics. No matter how exactly "fitness," say, is understood it is not semantically or cognitively on a par with "gravity" or "electric charge," since its meaning is not conveyed merely by means of mathematical laws and blind ostension. Thus the basic terms of biology express positive intelligible conceptions of things that we can either directly experience or can explicitly define, whereas in physics neither of these conditions holds. To put it differently, we have more than operational definitions in the case of psychology and biology—yet this is the sole descriptive content possessed by the characteristic terms of physics. Not surprisingly, then, physics is mathematics-dependent in a way that psychology and biology are not.[18]

Maybe molecular biology, being so close to chemistry, falls on the side of physics: but then it does not use the proprietary terms of biology and is simply a branch of chemistry. Genetics might be a hybrid of the two cases: part pure chemistry (structure

---

[17] Chapter 7 contains a detailed discussion of the epistemology of charge, wherein its mysterious character is emphasized. But biological function is not similarly mysterious, since we understand how an organ performing a function contributes to fitness—we know what it *is* to benefit the organism. Similarly, we know what the various organs of the body are—the eyes, the kidneys, the heart, etc. These concepts present their reference transparently, in the sense that possessing them enables us to say what the organ in question is—for instance the eye is the organ that enables the organism to see things (*eye* is a functional/teleological concept). Thus we know what eyes are in a way that we don't know what electrons are (*electron* is not a functional/teleological concept). The terms of physiology and anatomy are not I-know-not-what terms, with inscrutable reference.

[18] If we try to imagine physics without the formulation of mathematical laws, we come up with a pretty lame product, compared to physics as we know and love it. But in psychology and biology, as well as linguistics, history, and geography, it is possible to say quite interesting things without any mathematical treatment. Physics articulates its subject matter differently from these other disciplines, by relying upon mathematics; and this is an enormous part of its interest and charm. By contrast, the human sciences generate interest from the richness of their concepts, and their complex interrelations, not from precise mathematical relationships between remotely identified variables. Physics is thus beholden to mathematics for the substance of the discipline, while the human sciences are not—which in one way of looking at it is a count in favor of the human sciences. If scientists had all always been innate mathematical dunces, physics would look fairly shabby, but the human sciences would have pretty much the same interest as they have today. Maybe psychology would then have been the queen of the sciences!

of DNA and so on), part purposive biology (function of this gene or that). That's fine as far as my thesis is concerned, showing merely that the epistemological division I am suggesting cuts across conventional disciplinary boundaries. If genes have both biological functions, on the one hand, and electromagnetic properties at the molecular level, on the other, then they are part transparent and part opaque. In this respect, they are rather like the brain viewed in its ontological entirety: psychologically, the brain is transparent to us, since experience is an aspect of the brain's nature and it is given to introspection; but physically, the brain is as elusive as all matter, a congeries of electric charges and puzzling particles (as quantum theory assures us).

Neuroscience itself has two branches, insofar as it deals with the full reality of the brain: a science of experience as such, firmly grounded in intimate knowledge, and a science of the material basis of such experience, which shows all the opacity of matter in general. Indeed, experience and electricity, both salient properties of the brain, exhibit the duality I am insisting on with particular clarity: we know what the first is all right, but the second leaves us quite blank (except "structurally").[19] The brain is partly an object of intimate knowledge and partly an object of remote knowledge, since it is both psychological and physical. If we ask whether we know what a neuron is, the answer has a certain complexity: yes, in so far as neurons can be said to have a psychological aspect (we know our experiences and they are properties of neurons); and yes, in so far as they are biological entities with functions and conditions of well-being; but no, in so far as they are also objects for physics to describe and are subject to electromagnetic force, gravity, quantum effects, and so on. The neuron is simultaneously scrutable and inscrutable, conceivable and inconceivable—a curious hybrid of the phenomenal and the noumenal: a most peculiar entity. That befits its status as the crux of the psychophysical nexus.

What consequences can we draw from the split picture I am proposing? The first I have already anticipated: body is more of a mystery than mind, at least in one crucial respect. We simply don't know what body is—not really, not intrinsically. But we do know what mind is—really, intrinsically. The indisputable success of physics should not blind us to this elemental fact. We know the appearance of matter to our senses, to be sure, but physics has shown how far removed from the objective reality of matter this appearance is (modern physics is, above all, surpris-

---

[19] We have become so accustomed to the word "electricity" that we are apt to overlook the elusive character of its referent. It helps in regaining perspective to recall the state of knowledge when the science of electricity was first establishing itself (in the nineteenth century). Then it was clear to all that electricity (and magnetism) was a deeply puzzling, and rather counterintuitive, postulate—adventurous new frontier science, not the stuff of daily life (electric light bulbs, electric toasters, telephones). Conscious experience has never been similarly newfangled. The discovery of electricity was a startling new insight into nature, eerie, spooky, and controversial; but experience has never been anything like that. The truth is that we still don't know what electricity is, ultimately, though we have become very familiar with its powers and ways.

ing). The electromagnetic theory of matter demonstrates our ignorance of matter decisively, because of the remote status of our knowledge of electromagnetic force. The whole idea of field theory removes the essence of matter from the realm of perception, fields not even being perceptible. Particles dissolve into energy-gradients, which are grasped purely mathematically. Our grip on the physical is functional, which is to say purely by way of its effects, these consisting largely in measurements. But our knowledge of the mental is *not* purely functional—we know the nature of the things that *have* certain functional relations. Physics is, at bottom, a kind of black-box stimulus-response science, if I may put it so; but psychology deals with entities whose intrinsic nature is open to us (the irony here is rich, given much twentieth-century thinking about the mind). What sits inside the physicist's black box is an enigma to us, despite the mathematical regularity of its behavior (you put in the same stimulus, you always get the same response); but the mind is a transparent box with an open lid—we perceive its contents clearly.[20]

Does this make psychology (and linguistics and biology) *superior* to physics as a body of knowledge? In *one* sense, I am obliged to report, the answer must clearly be yes, since a body of purported knowledge is superior in proportion as it contains no areas of *ignorance*—and physics harbors large areas of ignorance (even concerning the things it is most authoritative about). It is disconcerting to be forced to conclude that our most prized science can't even tell us the real nature of its subject matter! But, of course, this wide valley of ignorance is coupled with supreme insight, because of the enormous success of the mathematical (and experimental) method in physics. We may not know what electric charge intrinsically is, but we know its mathematical and causal properties remarkably well. Indeed, it is amazing how thoroughly we know how electricity behaves given how ignorant we are of its nature. It is as if we were to know all about the proclivities of the person responsible for the crime, yet have no idea who that person may be.

Here we run up against what may fairly be called the paradox of physics: how it can be so informative and yet so uninformative at the same time? Psychology, by contrast, has the opposite profile: we know what it is about all right but we know relatively little of how the thing behaves—especially in terms of mathematical laws. And here we meet the paradox of psychology: how can we know so well what we are dealing with and yet know so little about how it works? Why is it so hard to find out the truth about something when it's so easy to know what that thing is? In other words, we have no trouble *interpreting* psychological statements, assigning clear

---

[20] The whole problem with twentieth-century psychology, in its behaviorist and functionalist period, was that it ignored and deplored the faculty of introspection, in a misguided attempt to make psychology "objective." But, I insist, introspection is a genuine source of knowledge, no less "objective" than anything else: it produces what Russell called knowledge by acquaintance, which is an especially valuable sort of knowledge. We have no such privileged route to knowledge in the case of physics, which is why physics has been so long in the making and is so fraught with difficulty—and so blind to the inherent character of its posits.

meaning and denotation to them, yet we find it difficult to make progress in discovering which psychological statements are true; while in physics the problem of interpretation seems endemic, yet we have good knowledge of what statements we are justified in asserting. Thus, famously, there is no agreed interpretation of quantum theory, despite its predictive success; but no one has a problem interpreting belief-desire psychology, weakly predictive though it may be.[21] Intelligibility and predictive success seem inversely related in the two cases.

The question of scientific realism looks rather different in the light of the dichotomy I am proposing. Let's by all means be realists in both cases, supposing that the terms of the two sciences denote real entities (with causal powers, etc.). Yet there is a sharp difference between the two: in one case we know what those real entities are, in the other we don't. The interesting case is physics: how can we be realists in good conscience about things whose nature we admittedly don't grasp? We might try saying that the only realist component of physics corresponds to its structural aspect, since that is something we know quite well ("structural realism"). But I think that assigns too much importance to our epistemic relation to the physical—as if only what we can know (the nature of) can be counted real.

I prefer to say simply that the basic terms of physics denote real things whose nature we don't know (and maybe can't know): thus matter is real (as real as mind) even if we can't know what it ultimately is (even if we can never know). Matter cannot itself be abstract, as Russell observes, even if our understanding of it is condemned to be; and its reality should not be made to depend on how accessible it is to us. Still, physics is shakier than psychology with respect to a realist interpretation of its content, and the pressure towards an antirealist interpretation is more intelligible (which is why radical operationalism or instrumentalism has seemed indicated to certain philosophers of physics). If we have to admit that we really don't know what electrons are, over and above the operations by means of which we detect them, or their relations to other particles, then believing in their independent existence can induce discomfort: but nothing like this obtains with respect to experiences. In physics we must make do with *agnostic* realism; in psychology we can affirm a more committed from of realism. In physics our realism must be *humble*, but in psychology we can afford to be *confident* realists.[22]

---

[21] Of course, there are different philosophical accounts of belief and desire, as there are of pain and visual perception. But in quantum physics the problems of interpretation are internal to the field, dividing physicists among themselves: these are not just philosophical differences. In a perfectly straightforward sense, we know what we are referring to with "belief" and "desire," even though we may differ about their analysis; but "electron" and "proton" are obscure even in the most basic sense in quantum theory, with widely different interpretations of what they might mean. (Wave-particle duality has no counterpart for belief and desire.)

[22] In case it is not clear, I do not claim here or elsewhere to have *argued* for realism in physics or psychology, or against various forms of antirealism (instrumentalism, etc.). I am merely stating my position and laying out the options and consequences. In fact, I am rather skeptical that there can be a successful argument for scientific realism, because the issue is too fundamental to admit of independent ground from which to argue. I merely record my conviction that it is really quite obvious

Here is the place to observe explicitly that my claim about the difference between the two types of science is purely epistemological. I am not at all saying that experiences are more real than electric charge or matter, or advocating anything even slightly idealist. From the point of view of an electron, it has a nature as transparent to it as the nature of experience is to us: if an electron could think, it would know just what being an electron consists in. It is not that charge is *inherently* less intelligible than experience—*de re*, so to speak. It just *happens* that we have access to the nature of experience, because of our special epistemic relation to it; we should not suppose that charge and matter and motion are somehow objectively opaque. God knows the nature of these things as well as he knows the nature of experience: but *we* don't. The mystery of matter is therefore a subjective or relative mystery (nothing is objectively mysterious). Even the deeply or necessarily unknowable has an intrinsic nature as robustly as the most easily known of things: epistemology never dictates ontology. So I am as realist about electric charge as I am about conscious experience. It is just that we are profoundly ignorant of physical reality, not that there is no reality to be ignorant of. My "structuralism" is epistemological not ontological.

Let me also be clear that in declaring matter to be a natural mystery, and mind not to be, I am not in any way retracting the "mysterian" position with respect to consciousness. The nature of experience is not a mystery, if by that we are referring to human knowledge of what experience is (its identity, we might say). What *is* mysterious, according to me, is how consciousness relates to matter.[23] Indeed, we can only appreciate *this* mystery if we already have a good idea of what consciousness intrinsically is: if we grasped it purely functionally, we would have no deep sense of mystery. So my view, in sum, is that matter is a mystery and the relation of matter to mind is a mystery, but mind is not itself a mystery (in the special sense intended here). Consciousness, I say, is a nonmysterious thing mysteriously related to a mystery. Perhaps the mystery of the relation owes something to the mystery of what it is related to: but that is a topic for another occasion.[24] Here my

---

that the sciences are attempting to discover what the ultimate constituents of reality are and how they operate—and that reality is indeed composed of hidden constituents with causal powers and objective intrinsic natures. Physical reality is in no sense a human construction. But of course I don't expect antirealists of various stripes to be convinced by this affirmation!

[23] See my *The Problem of Consciousness* (Basil Blackwell: Oxford, 1991) and *Consciousness and its Objects* (Oxford University Press: Oxford, 2004). The phrase "the mystery of consciousness" is systematically ambiguous between these two interpretations, and probably should be retired from discourse.

[24] The question is whether resolving the mystery of body would *thereby* resolve the mystery of mind's dependence on the body (the problem of emergence). Panpsychists think so, because body is shot through with mind, even before it forms brains. I don't share that optimism, but I think it is very likely that the mystery of body is part of the mystery of mind's dependence on body. Matter gives rise to mind somehow, and it does not do so by way of any attributes of matter that we know about—so presumably it does so by way of attributes we don't know about. Or else both matter and mind are expressions of some deeper reality whose nature wholly eludes us. Or...Whatever the case may be, mysteries cluster, so it seems a good bet that the mystery of body is connected to the mystery of its power to produce consciousness. (Of course, it can also be a mystery why something is a mystery.)

point is simply that it is not a mystery what someone is referring to when they use the word "consciousness"—or "experience of red" or "pain" or "thinking about home."

In case it appears mysterious why there should be this asymmetry in mysteriousness—with mind better known than matter—let me make an obvious point. Matter didn't strike us humans as all that strange when we were considering its middle-sized manifestations, with the sorts of velocities detectable by the naked eye, and without worrying too much about the nature of light: here the perceptual world seemed to offer a firm foundation for a general notion of matter. But that is a highly parochial perspective, and once we start to consider matter on other scales our familiar notions began to fray and collapse. Considering matter on the astronomical scale forces us to consider such things as the speed of light, excesses of gravitation, the possibility of cramming all of the universe's substance into a tiny atom-sized area at the time of the big bang—all things that put pressure on the reassuring perceptual world of middle-sized dry goods. But it is the small scale that really upsets the apple cart: the level of the atom overturns many of our cherished notions of how material things ought to behave, as twentieth century particle physics rammed home. None of this upheaval should be surprising: our old, comfortable notions of matter derived from our particular perspective on it; but that perspective is not privileged, and examining matter on smaller and larger scales was bound to produce surprises.

But nothing like this scale problem holds for the mind: there is no question of a biased size perspective here. It isn't surprising that the real nature of matter should have eluded us, given our particular skewed perceptual perspective on it—so the inscrutability here is predictable. But that is not our situation with respect to the mental (or the linguistic and biological): here we are at just the right scale to find out the truth—the important properties of minds (or meanings and organisms) are available to us from our normal perspective.[25] So it isn't mysterious why we grasp the important properties quite clearly in these cases. Not that we can easily discover how minds work—but at least we are not naturally disadvantaged in grasping the constitutive properties of mind. The basic ingredients of the physical world must be inferred from our less-than-ideal perspective on it, but the basic ingredients of the mind are laid before us in ordinary self-awareness.

---

[25] The problems that arise about mind and meaning—and they are very real—will not be solved by the invention of high-powered microscopes or by accelerating particles into the brain to check for diffraction patterns or by dissecting them into their tiniest parts (literally). We are not puzzled by mind and meaning because they are too small to see or are too far away or interact confusingly with our measuring instruments. In a way, the difficulties of doing physics are quite intelligible and predictable, given our observational vantage point; the difficulties of doing psychology and linguistics are puzzling by comparison. This is why we need special explanations of the intractability of mind and meaning (ordinary language is misleading, we haven't been at it long, it's all incredibly complex, etc.), while in physics the methodological obstacles are evident enough. Still, in the former case the difficulties clearly do not arise from simple cognitive remoteness—we are as *close* to mind and meaning as can be. The atom is not under our nose (*qua* atom), but mind and meaning are.

Reductionism may now take on a different complexion for us. Reductionism looks appealing if it can reduce the more puzzling to the less puzzling, the unknown to the known, the unintelligible to the intelligible: it has the virtue of eliminating areas of epistemic perplexity. But that motive lapses if the alleged reducing theory is more opaque than the theory to be reduced. Would we make experience any less perplexing by reducing it to electromagnetism? Not really—though we may thereby assimilate it to something more widespread and subject to rigorous science. But if you find yourself puzzled about what experience *is*, then you don't make much headway by reducing it to something whose nature is also unknown: you merely reduce one (alleged) mystery to another. There may be a point to that, but let's not deceive ourselves about ridding the world of all mysteries. Reducing everything to physics does not, contrary to the views of many philosophers, rid the cosmos of mystery: it just locates all the mysteries in one basket, where they stand forth in all their stark glory. If you believe that the aim of science and philosophy is to eliminate all areas of ignorance, so that everything becomes transparent to us, and you find the mind an area of mystery, then the *last* thing you should want to do is reduce everything to physics—the domain of mystery par excellence. Mystery, from my perspective, is something we have to live with, even in our most successful science, so there is no point in pretending to inhabit a universe in which nothing is hidden from us. Theoretical unification through reduction to physics may or may not be a good thing, but it is futile to seek it as a means of ensuring human omniscience.[26]

A standard argument against materialistic reduction is that knowledge of physical facts can never suffice for knowledge of mental facts (the "knowledge argument"). If mental facts were physical facts, then knowledge of the physical facts would entail knowledge of the mental facts: but it doesn't, so they aren't—so the argument goes.

---

[26] Idealism is the way to go if omniscience is your aim, because of the transparency of experience. If we can reduce the whole world to its experiential part, then we can render it accessible to our faculties of knowledge: it will be the *kind* of reality whose nature we can fathom. Instrumentalism is appealing for the same reason (and is in fact a form of idealism in its most consistent form). It seems to me no accident that idealism began its ascendancy at the time that the physical sciences were revealing how inaccessible the objective world is (Newton was pivotal here). The more we learned of the physical world, the more elusive it became—until matter itself became alien and inscrutable. Idealism seemed the perfect antidote: reality is really just human experience, variously ordered and arranged. Then there will be nothing real that demonstrates a deep human ignorance—nothing will be inherently incomprehensible to us. Idealism in all its forms—Berkeleyan, Hegelian, positivist, instrumentalist—is a protest against the unknownness of the physical world as science was revealing it. Its source is a fear of the radically unknowable. Idealism thus gained in traction the more science depicted the world as transcending human knowledge, out of a kind of epistemic desperation. Positivism was just a recent iteration. (Materialism, for its part, sheltered itself from the unknown by adopting a view of physics that avoided objective mysteries: but that ended up in idealist interpretations of physics, such as positivism and operationalism. So materialism turns into a form of idealism, taking its place alongside idealism proper, precisely in order to ensure that nothing about nature is mysterious. But mysterian physics is the correct position, adopted in effect by Locke, Hume, Newton, Kant, Hertz, Poincare, Eddington, Russell, and the modern structuralists. And if so, then reducing everything to physics does not eliminate mystery from the world.)

The considerations of the present paper support a different kind of knowledge argument, which I think is quite powerful. Namely, if mental facts were physical facts, then we would not know the intrinsic nature of mental facts: but we do, so they aren't. The reasoning is obvious: the entities of physics, particularly the forces, but also material substance itself, are unknown to us in their intrinsic character. So if facts of consciousness were simply identical to those entities, they also would have a nature that is unknown to us. But we do know the nature of conscious states. Therefore the materialistic reduction is unacceptable. Consider the ether, which was always conceded to have a mysterious nature, even by its supporters: if experiences were identified with ripples in the ether, then they would partake of the mystery of that supposed physical entity. But surely the nature of experience is not so remote from our cognition as that, *so* the ether cannot be the underlying constitutive nature of experience. Closer to actual reductive claims, suppose we try to reduce experiences to the electrical activity of neurons: then we run into the objection that electric charge has an unknown nature, being conceived only dispositionally, while experience does not, thus refuting the alleged identity. Thus "materialism" would make us more ignorant of the nature of experience than we actually are—as ignorant as we are of Eddington's "background" to the observable pointer readings.

But that is not our epistemic situation with respect to the mind: it is not an I-know-not-what. If body is a mystery and mind is not, then how can the mind *be* the body? In a certain sense, "materialism" is in the business, contrary to its intentions, of reducing the mundane to the exotic, the immediate to the remote, epistemologically speaking. And that gets things the wrong way round, since we don't find mind as such as puzzling as matter as such. The sensation of pain, say, is a lot clearer to us conceptually than the quantum-theoretic electron: so pain can't *consist* of such things. At the very least, if it does, we have to explain why the one *seems* so much more opaque than the other. How can sensations like pain be reducible to physical slithy toves, when talk of the former is so much more intelligible than talk of the latter? If we replaced one vocabulary with the other, we would go from clarity to obscurity, from the concrete to the abstract, from the categorical to the structural.[27]

---

[27] A philosopher like Russell simply takes it for granted that percepts (in his terminology) cannot be reduced to structural-mathematical features of things, because it seems so obvious to him that percepts have a full-blooded intrinsic nature that cannot be captured in the bloodless abstractions of physics, as he sees physics. Physics gives us only a "causal skeleton" of concrete reality, but percepts are fleshly realities—so they cannot be explained by the skeletal structure postulated by physics (though they may be reducible to the mental stuff that underlies the skeleton). The ontologically "thick" cannot be reduced to the ontologically "thin." On the contrary, the very thinness of physics needs to be supplemented by the thickness supplied by percepts, for Russell—so that matter can acquire some real concrete substance. The physical must be mental, not contrariwise. Russell has a robust view of the mental and a schematic view of the physical, so the idea of reducing the former to the latter seems out of the question to him. Whatever one may think of this position, it is surely a deeper and more sensitive assessment of the situation than the usual kinds of uncritical "physicalism" we have been subjected to in the years since Russell wrote.

Let me make explicit what must be (perhaps annoyingly) plain to some of my readers: I am rejecting verificationist or operationalist or instrumentalist views of physical theories, as well as behaviorist views of mental concepts. If we were to claim that there is simply no referential content to physical theories beyond the meter readings and other observations associated with them, then of course there would be nothing spoken of in physics that exceeds what we can know. I am, to the contrary, assuming that these observable matters are merely indicators of an underlying (transcendent) reality. I have not defended that view here. I am also assuming that our first-person access to mental states affords us unique insight into their nature—which, again, is a view not shared by all philosophers. My arguments result from accepting these two assumptions—agnostic realism in physics and first-person authority in psychology—and seeing where they lead. That is, we know matter only by, and in terms of, its observable symptoms, which are remote from its objective reality; but we know mind directly, without mediation by its symptoms.

If we knew mind only from a third-person perspective, then our knowledge of it would be as remote as our knowledge of matter—we would think of mental states only as "that which lies behind this behavior, whatever exactly that might be." This is essentially how we think of matter when we are doing physics: particles, fields, and forces are just "that which lies behind these measures, whatever exactly that might be." And it is because of this lack of specificity that various metaphysical conceptions can step into the breach—claiming variously that matter is really mind-stuff, or is some sort of impersonal will, or is something neutral between mind and matter-as-perceived ("neutral monism"), or consists merely of deformations of space, or of gaps in space, or is just brute dispositions to motion, or is what happens to super-intense heat when it cools down (big bang cosmology), or is constituted of vibrations in the void (string theory as metaphysics), or even consists of varyingly pitched notes of pure music.

The descriptive content of physical theory, when such theory is interpreted realistically, simply leaves open a variety of options for characterizing the objective entities whose behavior the theory details. It tells us that "something is doing something" but it doesn't pin down what that something is: it gets content only by dint of the mathematical relations it discerns in the measured phenomena. But, I have insisted, that is not our epistemic relation to states of mind: they are not mere posits, identified from afar. At any rate, realism in physics and first-person access in psychology are my assumptions here. Still, I think these are solid assumptions, and they lead to what strike me as sensible conclusions about physics and psychology and their interrelations (though they will no doubt strike others as outrageous). As so often in philosophy, we must look at the whole package.

All would be different if a thrilling conjecture of Eddington and Russell were correct, namely that by having a brain we gain access to the intrinsic nature of matter. Then the nature of electrons and other particles would turn out to be experiential in character. That would deliver the result that we *do* know the nature of matter, deeply, intrinsically—at least at one of its locations (in brains). This nature is straightforwardly mental. Then the mystery of matter would be solved: behind the meter readings, lurking in the "background," lies pure consciousness—and pure consciousness all the way down to the most basic physical realities (as if those vibrating strings were just the *hearing* of musical notes). But I am not, to put it mildly, convinced by this exciting and ingenious proposal, so I prefer to adopt an agnostic position regarding what might be the ultimate nature of matter.[28] There would, on the Eddington-Russell view, be in the end only *one* type of science—the type of psychology—since all physical facts would bottom out as psychic facts. Physics would be, in effect, the psychology of the cosmos. But if you find that position unappealing, as I do, then you are left with the dichotomous picture I am defending: psychology deals with facts into whose nature we have nonnegotiable insight, while physics deal with facts that transcend our powers of comprehension.

Finally, where does philosophy stand in this big scheme? Philosophy is not usually regarded as a science, but it can hardly be denied that it has a subject matter of some sort—so what is its epistemic relation to its subject matter? Is it like physics, identifying its subject matter remotely, silent about the intrinsic nature of that subject matter? Or is it like psychology, quite clear about the nature of its domain, if not terribly impressive in the results it has achieved? Well, that depends what we take philosophy to be *about*—what its organizing concept is. What concept stands to philosophy as *force* does to physics, as *purpose* does to biology, as *mind* does to psychology? What is philosophy the "science" *of*?

---

[28] I discuss panpsychism in *The Character of Mind* (Oxford University Press: Oxford, 1996; first published in 1982), 33–36. I would add here that the laws of physics are not recognizable as psychological laws. If matter were constituted by mental states with causal powers, even in part, then we would expect the behavior of matter to exemplify psychological laws—but that seems flat wrong. Suppose, for simplicity, that the mental states were of the nature of beliefs and desires (perhaps "proto" beliefs and desires): then particles would act as if they had a belief-desire psychology. But they manifestly do not. The physical universe has no *psychology*. The only response the panpsychist could make to this point is to detach the possession of mental states from the possession of a psychology—in the sense that the system exemplifies psychological laws or dispositions. But that is a highly implausible maneuver: what sense does it make to attribute mental states to particles (and cells and organs) if there are no consequences for the dispositions of the entities in question? I say: "no psychological states without a psychology to go with them." One response might be that the mental states of subpersonal entities are so alien to us that we cannot recognize a psychology in the behavior of those entities, though there is one. That is surely desperately ad hoc and implausible. Another response might be that the entities *do* manifest a recognizable psychology, so that the link between psychological states and a psychology has not been broken: thus an electron shows its desire to be near protons by attracting them, and its wish to avoid other electrons by repelling them. I take it this proposal is so absurd as not to need refutation. The simple fact is that only whole organisms act in such a way as to manifest a psychology, and so only they can be credited with psychological states.

I think there is still no better answer to this meta-philosophical question than the following: *concepts*.[29] Philosophy is about concepts, as physics is about force, as biology is about purpose, as psychology is about mind. It may be about concepts in a different *way* from the way those other sciences are about their proprietary entities—being an a priori not an a posteriori investigation—but it is *about* those things in just as robust a way as the other sciences are about their chosen entities. Philosophy wants to know the nature, laws, and inner constitution of concepts: what they are, how they work, what their specific content is (I have been talking here of the difference between physical concepts, on the one hand, and mental, linguistic and biological concepts, on the other). The question then is what the epistemic relation is between concepts and philosophical thinkers: is it like the relation between physicists and electrons or is it like the relation between psychologists and thoughts or sensations? Do we philosophers have remote knowledge of concepts or intimate knowledge?

The question is not perhaps as easy as it looks to answer. On the one hand, concepts are plausibly taken to be psychological entities—the contents of thoughts, after all—so we ought to be well acquainted with their mode of being. We are scarcely peering at them from afar, through a distorting lens, mediated by ingenious measuring devices. Yet, when it comes to excavating their inner constituents, or even their general nature, concepts can be remarkably recalcitrant, stubbornly opaque. We simply don't have automatic reflective knowledge of how our concepts should be analyzed. Nevertheless, I think it is clear that they belong on the side of psychology not physics: for they *can* be analyzed, and when they are, their inner architecture is revealed. It is not that we remain in the dark about their real character once an analysis has been completed, as we do when we have completed our physical theory of electromagnetism. There is no mysterious residue once the analysis has been carried out, and it is possible to complete the analysis—at least in principle and in some cases.[30] I thus assert that concepts are the *kind* of thing that can be transparent to us, unlike the basic quantities of physics: we really can get at their nature, not merely their remote symptoms. A theory of concepts is not condemned to mathematical abstractness, to slithy tovehood, to a maddening elusiveness. Accordingly, philosophy is not only possible, it is capable of yielding *intimate* knowledge of its subject matter—real intelligible insight into what it is about, viz.

---

[29] Again, not everyone will agree with this account of the nature and subject matter of philosophy. I defend it in my *Truth By Analysis: Games, Names, and Philosophy* (Oxford University Press: New York, 2011).

[30] There are trivial examples such as "husbands are married males," but also far-from-trivial examples like Bernard Suits' analysis of the concept of a game: see his brilliant book *The Grasshopper: Games, Life and Utopia* (Broadview Press: Ontario, 2005; originally published in 1978). My point is that in such cases we do gain real insight into the makeup of our concepts: we are not confined in our knowledge to knowledge of effects or merely structural features. We thus acquire intimate knowledge of the composition of our own thoughts.

concepts. In this respect, at least, it shares with psychology, its close cousin, the notable merit of outshining physics as a vehicle of knowledge into how the world really is. Philosophy can tell us how certain things (concepts) intrinsically are, in a way that physics cannot with its defining entities. Physics is like a blind genius; philosophy is more akin to a sighted mediocrity. Both have their respective virtues and shortcomings.[31]

---

[31] Are there any grounds here for "philosophy envy" on the part of physicists? I am not entirely joking when I ask this impertinent question: for I do think that philosophy has a genuine claim to supreme epistemological standing—unfashionable and shocking as it might be to say such a thing today (see the book cited in note 29). Philosophical knowledge has certain virtues that no other knowledge possesses, I think, particularly in the way of thorough and definitive understanding—though it is obviously hard to acquire such knowledge. This is precisely why so many great minds, from Plato on, have sought philosophical knowledge, and been happy with nothing less. (What I say here is compatible with admitting that little or no philosophical knowledge has been acquired: I am speaking of the ideal.)

9 The Ontology of Energy

The concept of energy has not always played the central role in the physical sciences that it does today. The notion of corpuscular material substance extended in space was taken as fundamental, beginning with the Greek atomists. In Newton's *Principia* the concept of energy does not even appear (though there much about force). But recent physics has accorded energy a position of highest honor—as *the* basic physical reality. The universe is now generally conceived as constituted by energy, or at any rate pervaded by it, with matter as just one form of energy. As the idea of a field became more central (and respectable), it became natural to speak of energy as embodied in a field, and hence as distributed in space, as evidenced by the existence of a force at points of space.

Einstein's famous equation $e=mc^2$ announces the preeminence of energy. However, it is not terribly clear what energy *is*: when we attribute energy to a system *what* is it that we are attributing? What is the ontological status of energy? What is its objective nature? How exactly is it related to such things as heat, motion, electricity, mass, and gravity? When an object possesses a quantity of energy, how does it differ from another that lacks that energy? Some earlier theorists, of idealist sympathies, took it to consist of something like will, possibly of the divine variety, no doubt anxious to find some familiar reality with which to understand energy. But eschewing such fancies, there is a question as to what energy really consists in—about its ontological standing. If our physical ontology is to include energy, what exactly is it including? And how robust a reality is energy? Physicists routinely employ the definite description "the energy of system S": what exactly are they referring to, if anything? The physics textbooks tend to pass quickly to scales of measurement and experiments, but the metaphysical question cannot be avoided. What kind of *being* does energy have?

I propose to explore this question by considering the *conservation* of energy across different physical systems over time. We know on good authority that energy is conserved and that this conserved quantity can take different forms; so we can ask *what* it is that is thus conserved—that will be what energy is. Electromagnetic energy can be converted to chemical energy (by photosynthesis), and this in turn can be converted to kinetic energy (as with the movements of animals)—all the while maintaining the original amount of energy. Energy cannot be lost, and it also cannot be gained (so the big bang cannot actually have created any energy): it is a constant. It can appear in one form and then reappear, undiminished, in another form. Thus energy is (apparently) *transferred* from one thing or system or spatial point to another—communicated, passed on. As it does so, it is transformed, embedded in a new type of physical substrate; yet it remains constant. The energy in a gravitational field, say, is transmitted to the motion of massive objects, which can then pass it on to other bodies. Thus ontological diversity in the underlying entities is consistent with a fundamental sameness in energy content. Over time an imperishable something hops from one physical location to another. The universe changes while also remaining the same.[1]

My question is how to interpret these facts of physics—what should the ontology of energy look like? There are two basic theories: conservation is the identity of a *measure* or it is the identity of a *thing*. According to the first type of theory, it is merely that we make measurements that turn out the same as energy is (said to be) transformed, and that is all that the conservation of energy amounts to. If we measure energy by means of *joules*, then energy is just the measured joules in a system, i.e., a certain number. According to the second type of theory, however, there is an underlying entity or material or stuff or substance that is physically continuous throughout the process of transformation—a thing that *grounds* the constancy of measurement. Energy, then, is not a number, but an objective physical reality: we

---

[1] This, at any rate, is the natural way to put it, though ontological deflationists about energy will find such a formulation tendentious. If we introduce a type-token distinction with respect to energy conservation, then we can say both that the *property* (type) of having such and such a quantity of energy remains constant and that individual *particulars* (tokens) consisting of energy packets persist unchanged over time. Type energy conservation is consistent with new tokens coming into existence over time whenever energy is "transferred"; but token energy conservation insists that the very same individual token of energy is preserved when energy passes from one system to another. In the latter, stronger, version, concrete particulars, clothed in different energy forms, enjoy an immortal existence, because nothing can destroy them—they are rather like eternal particles. Token energy conservation treats energy as consisting of individual substances that endure, but type energy conservation merely says that energy as a general characteristic does not vary (so the type persists, though not the token). The relation expressed by "same energy" can then either take tokens or types as its relata, as when we say, "system *S1* has the same energy as system *S2*." As will emerge, I accept both sorts of conservation.

assign a number to this reality, according to a chosen scale, such as the joule, but it is not itself a number.

The usual definition of energy in physics does not suffice to decide between these two theories, but it is still a useful place to start. Energy is most generally understood as the capacity to effect change, but in physics the idea of change is usually specialized to "work," where work is taken to involve motion. This is by no means semantically required by the word "energy": it reflects the focus on motion that is characteristic of physics as we have it.[2] Let us, however, follow physics and define energy simply as the capacity to generate motion. Then we can say that conservation entails that the capacity to generate motion remains constant. This capacity may shift location, but it does not vary in amount. The question accordingly becomes: when we speak of the capacity to generate motion, do we mean a measure or a thing? Is it merely that the capacity in question is measured as the same over time, or is it rather that the capacity springs from a constancy in physical reality?

The question was debated by philosophically minded physicists in the early days of energy theory. Lange quotes Heaviside:

The principle of the continuity of energy is a special form of that of its conservation. In the ordinary understanding of the conservation principle it is the integral amount of energy that is conserved, and nothing is said about its distribution or its motion. This involves continuity of existence in time, but not necessarily in space also. But if we can localize energy definitely in space, then we are bound to ask how energy gets from place to place. If it possessed continuity in time only, it might go out of existence at one place and come into existence simultaneously at another. This is sufficient for its conservation. This view, however, does not recommend itself. The alternative is to assert continuity in space also, and to enunciate the principle thus: When energy goes from place to place, it traverses the intermediate space (137–8).[3]

Commenting on Heaviside's evident realism about energy, Lodge says:

This notion is, as I say, an extension of the principle of the conservation of energy. The conservation of energy was satisfied by the total quantity remaining unaltered; there was no individuality about it: one form might die out, provided another form simultaneously appeared elsewhere in total quantity. On the new plan we may label a bit of energy and trace its motion and change of form, just as we may ticket a piece of matter so as to identify it in other places under other conditions; and the route of the energy may be discussed with the same certainty that its existence was continuous as would be felt in discussing the route of some lost luggage which has turned up at a distant station in however battered and transformed a condition. (138)

---

[2] I discuss physics and motion in chapters 3 and 4.

[3] Marc Lange, *An Introduction to the Philosophy of Physics: Locality, Fields, Energy, and Mass* (Basil Blackwell: Oxford, 2002). Page references occur in the text.

Thus conservation is not conceived merely as invariance in the results of measurement, but as the actual continuity over time of the self-same entity. By contrast to such realism, Jeans impatiently asserts: "The mathematician brings the whole problem back to reality by treating this flow of energy as a mere mathematical abstraction" (148). We may speak naively about the transfer or flow or continuity of energy, as if this was the motion of a subtle substance through space and time, but in fact all we have is an invariant numerical value, for Jeans. Heaviside and Lodge reify, while Jeans reduces (or eliminates). One side takes "energy" to refer to a persistent stuff, while the other thinks of it by analogy with "the average man."[4]

We must not let the customary operationalism of physics, as it is typically expounded, dictate the answer to our question. We should not suppose that since *everything* in physics is to be defined by how it is measured, energy is just a special case. That would rob the question of its specific interest. For in the case of energy there seem to be distinctive reasons to adopt an operational interpretation, since energy is so closely tied to *doing*—energy is what makes things *act* in certain ways (perform "work"). So the question is whether the conservation of energy just consists in the fact that things keep acting in the same way—performing the same amount of "work." The total quantity of *action* in the universe is what is invariant—and action is not a stuff or substance or entity. There is no reduction or increase in the overall amount of action the universe contains, i.e., movement (actual or potential): and to say this is not yet to say that any underlying *entity* persists through space and time—it is purely a statement about what happens at the observational level. So energy might be said to be uniquely operational: it just consists in kinetic facts—propensities to motion.[5] And this might be thought to favor the metrical conception over the metaphysical conception. So, again: is the energy of a system just a measure of its ability to do work or is it a *sui generis* physical reality? Is there some physical reality that is measured or is the measuring all there is?

---

[4] Here is where the deflationist or antirealist about energy will insist that all talk of energy transfer is just a "useful fiction," not a report of hard fact—nothing real is really transferred in reality. The similarity to other ontological questions is evident: fictionalism about numbers, substances, theoretical entities, moral values, modality, causality, mental states, space, and so (drearily) on. That average man has a lot to answer for. What is curious, of course, is that our discourse concerning these things is nothing like real fiction—invented by novelists and the like for our entertainment and edification.

[5] I am here trading on the usual ambiguity in operationalism: between the thesis that physical quantities should be defined in terms of measurement (the strict reading), and the thesis that we should focus on the event, process, or action and not on the object, substance, or entity (the loose reading). On the former reading, the relevant operation is performed by the observer, as he or she measures physical reality; on the latter, the operation is performed by reality itself, as it goes about its active business. These are very different ideas, though commonly conflated. The two are connected through the phenomenon of motion: for motion is both what matter does and the prime object of physical measurement. To focus on motion is to focus on the active *and* the measureable. Operationalism about electric charge, say, will then consist of the conjunctive claim that charge is exhaustively expressed through the activity of motion and that it is best defined in terms of the measurements of motion that we make when we seek to measure charge.

I want to approach this problem by comparing it with some standard positions on the nature of mental states. Consider the different forms of energy there are: chemical, electrical, kinetic, gravitational, elastic, etc. Each of these carries a particular energy content, which remains constant as the energy is "transferred" from one medium to another. The kinetic energy of an electrically driven fan, say, is identical to the energy contained in the electrical current that drives it (assuming no leakage). Now we can make a three-way distinction: first, there is the medium itself, the particular physical material in question (electrical or kinetic in our example); second, there is the energy that is contained in that medium, that which is conserved across media; third, there is the functional or dispositional property of having (or being prone to have) a certain effect—as it might be, making the fan blade rotate.

The question then is what the relation is between these three concepts: do any of them denote the same thing as the others? The medium, the energy, and the disposition to cause motion: are any of these the same? Compare: a brain state, a mental state, and a functional property such as the power to bring about a certain bodily movement. According to a familiar story, we can't identify the mental state with the brain state, because there can be many physical bases for the same mental state ("multiple realization"). This has encouraged some theorists to identify it with the functional property—while still others have insisted on a distinction here too. The properties are clearly closely interrelated, but it is not so clear that any identities hold.

A similar dialectic can be described for the case of energy: is the energy a specific state of the medium, or is it the functional property of doing certain work, or is it distinct from both these things? It can't be the first, apparently, because the *same* energy can be contained in different kinds of physical system, as when it is shared by an electrical system and a kinetic system: the energy can be "multiply realized." But the functional property does not vary in this way—it is what is in common between the different systems in just the way the energy is. The functional property is simply the power to carry out work, i.e., produce a specific state of motion—and that property can be possessed by different sorts of material substrate. So should we be "functionalists" about energy? Is energy just the power or disposition or tendency to do work? To be sure, this functional property has a *basis* in nonfunctional physical facts or conditions, such as an electrical current or a set of motions; but it is clearly not the same as those—so it seems to exist at the right level of generality to be the energy that is conserved. The functional property is preserved under variations in the medium, just like the energy, and it shares with energy a direct connection with the concept of action.

Put simply, then, energy is the power to act (not the specific mechanism of the action), and so the conservation law states that this power remains constant. It is not, on this view, that energy *has* the power to generate action—as mental states (apparently) have the power to produce behavior—while being ontologically distinct from that power: rather, the energy collapses into the power. Thus, there is just the physical realization in the medium and the associated functional property—there is no entity poised, so to speak, between them: the energy itself.

This is no doubt a tempting position, what with the principle of parsimony and so on, but I think it is wrong ultimately. The trouble is that having the functional property in question is not logically sufficient for a system to possess energy. The easiest way to see this is to consider a world in which the active powers of objects in the actual world are replaced by passive powers in the objects they interact with. Instead of electricity having the active power to generate motion in objects, objects have the passive power of undergoing motion when contacted by electricity: electric charge doesn't *make* motion happen; it merely is the occasion of motion in the "receiving" object. Normally, we say that a body in motion has kinetic energy and then communicates that energy to another body by collision: well, let us consider a world in which there is no such energy in the first object, but the second responds *as if* the first has energy, because of *its* passive powers. In such a world, the first object has the functional property of being counterfactually connected to the second object in the appropriate way, but no energy is going *from* the first object *to* the second. The first object does not have the kind of active power that possessing energy confers; yet it is still true that if the first object collides with the second, a certain change in the motion of the second object occurs. For example, imagine colliding billiard balls in which the motion in the target ball occurs because of a tendency in that ball alone to move when the cue ball touches it, but the cue ball imparts no energy to the target ball. In such a scenario the cue ball has the following functional property: being counterfactually connected thus and so to the motion of the target ball. But there is no transfer of energy from cue ball to target ball, just a certain type of counterfactual dependence. The problem is the familiar one that counterfactual notions cannot capture categorical notions—in this case, the notion of possessing and transferring energy. That notion presupposes an attribute or state of the energetic object that is transferred to another object, but the merely functional view cannot make room for this idea of transfer. Or again, can't we conceive of a deity that ensures that the right effects occur, with work done being conserved, without any energy in things being *responsible* for this (as with Occasionalism)? The thing responsible for the effects is the deity's intervention, not an attribute or state of energy possessed by the objects in question. A world in which objects *had* no (intrinsic) energy could still be a world in which the counterfactuals that define the functional property obtain, but in virtue of the supervising deity's externally imposed actions.[6]

---

[6] I am obviously thinking here of Leibnizian pre-established harmony, in which no causation proper holds between worldly objects or events, despite appearances, because God has set things up so as to give the false impression of causality. The counterfactuals could all remain true, but their categorical ground would not be causation-in-the-world but God's original (or repeated) act(s). In the same way, God could set things up counterfactually to give the impression of energy transfer, without there being any such thing. If so, the holding of the relevant counterfactuals cannot be logically sufficient for energy transfer to occur. The truth is that the counterfactuals obtain in virtue of the actual transfer of energy, not that the transfer occurs in virtue of the holding of the counterfactuals. This kind of problem is endemic to all sorts of counterfactual analyses of (seemingly) categorical notions.

It might be said in reply that the functional property is stronger than mere counterfactual dependence; it includes the idea of an *ability* to do work. Could an object or system have the ability to do work and yet possess no intrinsic energy? Well, if we mean that it possesses an attribute in virtue of which it can produce such and such effects, and that attribute is the ability, then this is not really a functional view—it is tantamount to the idea that energy is something that underlies the ability, and gives rise to it. Energy is what *gives* objects the ability to change other objects, but the ability cannot just exist on its own without a categorical basis. The ability cannot be "bare." Of course, an object with energy has a certain ability to do work: but the question is whether the ability can exist without anything categorical to ground it.

We can express the point in another way. Suppose we say that the work done in the universe is a constant—that action is conserved. We have discovered that there is no decrease or increase in how much motion can be produced by the universe over time. Then the question arises as to *why* that is so: what *explains* this constancy at the level of observations? If the conservation of energy amounted to no more than equality in work done over time, then we cannot genuinely explain this constancy by appeal to the conservation of energy—since that would be the *same* fact as the fact purportedly explained. But if we think of energy as a *bona fide* entity that remains the same over time, then we *can* say why the universe behaves in this unvarying way: it is because energy—the underlying thing—remains the same. The capacity for action is constant *because* there is a persisting distinct reality called energy. There is literally a transfer of energy stuff from one system to another—an entity shifting location—and that is *why* the two systems generate the same work. This view deserves to be called a *realist* view of energy, because it makes energy transcend our evidence for it. If we could not provide such an explanation, on logical or analytic grounds, then it would really be just an *accident* that work done remains invariant, since no feature of the universe would explain it. But if we accept that a certain entity exists invariantly over time, and is distinct from its various effects, then we can *predict* that work done will be a constant. *Given* that energy is an imperishable physical reality, the universe will naturally act in such a way that work done remains constant. Why is there this observed constancy at the level of action? It is because at the level of underlying entities there is something there that persists. Energy therefore *causes* work to be done; it does not *consist in* work done. Logically, it is just like the case of mental states: mental states cause behavior; they aren't constituted by behavior. Mental states do indeed have functional properties, but they do not consist in functional properties—do not reduce to them. Energy likewise has functional properties, but it is not identical to a functional property—it is not so reducible.

In the case of energy, though, we have in addition the conservation law, and that tells us that something is passed between objects, so that different systems can be forms of the same energy. The systems behave in the same way, as measured by their capacity for work, precisely because they have a physical attribute in common—the energy they contain. Why does a given billiard ball have the same kinetic energy as the ball that just collided with it? It is because something has literally been transferred from one ball to the other. If the conservation law just amounted to the statement that the two balls in fact happen (as a matter of law) to have the same kinetic energy, then that would be a kind of cosmic coincidence. But the law is generally stated in the more ontologically committed fashion that I am recommending: that a certain physical quantity is transferred from one object to the other, flows from one to the other, and nothing is lost in the transfer. The energy in one has the same numerical value as the energy in the other because it is the *same* energy existing in both: numerical identity underlies numerical identity, we might say.[7] Just as atoms are supposed to persist unchanged as they shift from one location to another, forming parts of distinct objects, so units of energy remain what they are as they move around the world in different energetic systems. That, at any rate, is the realist view I am suggesting. The *reason* energy is measured to be the same over time is that it is the same *thing* over time: the *explanation* of measurement constancy is ontological constancy.

This may seem like a strange position to some, a gratuitous reification of something that ought to be a mere mathematical pattern. Couldn't we at least take some of the ontological sting away by identifying the energy with the form it takes in any given system? Can't kinetic energy just *be* motion and electrical energy just *be* electric charge? Why do we need an extra ontology of energy over and above the physical medium in which energy "resides"? I don't think this reductive view will do, however, because it is *contingent* that a particular motion or atomic configuration should have the energy that it does. We can conceive of the motion occurring without the energy or the configuration existing without giving off any energetic power. Having kinetic energy is not the *same thing* as simply moving. The truth is that the motion or configuration *contains* a certain amount of energy, without itself *being* that energy. The energy is something *additional* to merely moving at a certain rate or being made of certain chemicals. It is ontologically separable.[8] This is precisely *why* energy is conserved but the various forms of energy are not. The conservation of energy law doesn't state that the amount of a specific form of energy stays constant; it states that the energy contained *in* specific media stays constant. Specific media could go out of existence altogether, so far as the general conservation of energy is concerned; indeed, all energy could be converted ultimately into homogeneous kinetic energy, consistently with the conservation of energy in general. Energy is something over

---

[7] Just to be perfectly clear: strict (numerical) identity of entity underlies identity of numerical value—same thing, same number.

[8] This is why physicists speak of the *same* energy being contained in *different* systems (we could mean type or token here): the energy is common to both, yet the two systems are physically dissimilar. The multiple realization of energy (same energy, different physical basis) precludes any case-by-case reduction. Compare the claim that pain is C-fiber firing in humans, X-fiber firing in Martians, Z-oscillation of microtubules in Venusians, and so on: that is not a satisfactory identity theory of pain (the type), by the elementary logic of identity.

and above electrical fields or motions of bodies or chemical bonds: it is what those things contain and can pass on. Objects are said to *store* energy or to *accumulate* energy, and what is thus stored or accumulated is not reducible to the medium of storage. When energy is released the object loses something substantial that it once possessed, which now enters into another object.[9]

I conclude, then, that energy is a single homogeneous thing that takes up residence in different material forms—light, heat, tension, masses, etc. It is a real attribute or state that different types of physical system can share. The conservation of energy law therefore reports the persistent existence of this entity, and its spatial transference from one system to another. It is capable of dramatic transformation, appearing now in one form now in another, but it runs through a variety of phenomena, always uniformly itself. Thus the usual talk of the transfer or flow of energy should be taken at face value and not watered down in the operationalist style. It is not that this way of talking is metaphysically misleading, fit only to be replaced by talk of the constancy of work done or of measurements performed. Energy is an imperishable physical reality that passes between objects or systems, never losing any of its intensity (or gaining either).

I have spoken so far only of physical energy, but there is also the notion of mental energy. Whether this form of energy is reducible to the physical forms—say, electrical potentials in the brain—is a difficult question, turning upon the general answer to the mind-body problem. But even if it is not so reducible, so that mental energy is a *sui generis* form of energy, there is still, according to the present position, a deep commonality between the mind and the rest of nature—because all energy is fundamentally continuous. The various forms of energy all share the same underlying continuous reality, and this is as true of mental energy as the rest.

Assuming that mental energy is also conserved and transferable, the very stuff that constitutes it can constitute other forms of energy, say the kinetic kind (for example, the energy involved in thinking can be converted into the movements of a pen). Physical energy can be converted to mental energy and vice versa, so that a common thread runs through the mental and physical worlds. The energy of an emotional reaction, say, could be the very same energy that once fuelled the growth of a plant. Thus we have a monism of sorts, and it is compatible with even extreme forms of dualism: after all, magnetic fields and bodies in motion are very different things, but

---

[9] There is a supervenience question here: is energy content supervenient on the physical features of a system (specified without mentioning energy itself)? It is not clear what to say here: did God need to do more work to create energy than merely to create the physical systems in which it resides? Can we have "zombie energy worlds" in which the same physical systems exist but contain no energy? My sense is that we can, because energy content is genuinely something over and above ordinary physical properties, such as motion or intra-molecular bonding: but the question is difficult to answer. Certainly, energy content is a different *property* from the physical properties of the underlying system. (It is interesting how obscure the ontology of energy is, given its vital role in modern physics.)

they can share the same energy content. The energy involved in thinking can be the very same thing as the energy involved in moving objects around, since all types of energy are inter-convertible—but this is consistent with thinking not itself being a type of motion. In a certain sense, then, the energy of the will *is* identical to the energy found in physical nature—even if the will itself is not at all like other forms of energy. Still, the uniformity of energy across different domains demonstrates that the mind is not as cut off from the rest of nature as some philosophers have assumed (I am thinking of Descartes and the religious idea of the soul). The mind shares its ability to effect change with the energetic force of the rest of nature: what powers the processes of thought is identical to what powers everything else, i.e., universal energy. We can assert an *identity* theory of mental energy and physical energy.[10]

I have suggested that energy is a real objective trait of reality, not merely a useful mathematical fiction. I have also suggested that it has a highly general nature, which is capable of assuming different forms. But I have said nothing about what its intrinsic nature might be. Not surprisingly, given the general trend of this book, I incline towards a position of agnostic realism about energy: we refer to something we-know-not-what with the word "energy"—just as we do with the word "matter." We know that energy circulates and remains constant, as well as knowing the laws of its various manifestations, but we don't know what it consists in—we have no positive descriptive conception of it. Given that it is basic to physical reality, then, we lack knowledge of a fundamental aspect of the world. We know its abstract mathematical character, and its role in empirical theories, but our knowledge does not penetrate to its underlying essence. We use the word "energy" to designate a theoretically useful enigma. In this the term belongs with the other theoretical terms of physics. Physics based on the concept of energy is therefore no more ontologically perspicuous than other forms of conceptualization (for example, those based on the concept *matter*).

---

[10] The energy of a particular thought or feeling can thus be identified with the energy of the correlated neural state, which is primarily electrical energy. The ability of the thought or feeling to do "work" is grounded in the ability of a neural state to produce other neural states or bodily motions. Mental energy is really strictly identical to neural energy. This aspect of the mind is then reducible to an aspect of the brain: it is literally true that electrical energy is the cause of the effects brought about by thoughts and feelings. Energy passes from the external physical world to the brain, enters the mind, and then produces mental "work." None of this seems remotely compatible with Cartesian dualism, and certainly links the mind firmly to the brain, where it belongs. Yet there is no entailment from this point about mental energy to any general reduction of mind to brain: states of mind themselves could be irreducible to brain states compatibly with their energy content being so reducible. Consciousness itself, though irreducible to (the usual) brain states, still can be said to have its energy content identified with brain energy (electricity, so far as we know). Have we here stumbled across the grain of truth in old-style identity theories?

10 Consciousness as a Form of Matter

Cartesian materialism is the doctrine that everything about the physical world consists in, and can be explained by, extended massive bodies interacting by means of contact causation. Descartes held this view about the physical world, but he rejected it for mind. The view is sometimes called "mechanism," or characterized as the doctrine that nature is a "machine," but these terms are somewhat too inclusive to capture the intended doctrine, because there seems no obvious reason why a mechanism or machine has to work purely by contact causation. Couldn't there be a man-made machine that operated by means of action-at-a-distance—indeed, don't many machines exploit gravitational force?

In any case, it has turned out that Cartesian materialism, as I prefer to call it, is false. First, there was Newton's appeal to the "occult" force of gravity that operated mysteriously through a vacuum. Then we had the shock of Maxwell's theory of electricity and magnetism, which postulated an "immaterial" field as the medium of attraction and repulsion between charged bodies. Galileo tried to vindicate Cartesian materialism with the idea that gravity was caused by a downward wind, but eventually it had to be accepted that the old model of clashing atoms in the void had to be replaced. The idea of the ether was another failed attempt to rescue gravity from occult status. Thus the era of the field of force was upon us—spread out in space, falling off with distance, boundary-less and devoid of surface, overlapping and penetrable. The notion of material things as sensible (in both senses) solid bounded entities that transmit force by collision had to make room for a broader physical reality that included invisible lines of force pervading the vacuum. And this was all before the anomalies and paradoxes of quantum theory. If we restrict the term "matter" to the formerly accepted range of entities, then we had to admit that our ontology must include things other than matter (even before we get mind). In fact, that term, as well as "physical" and its cognates, was widened to take in the new realities required by theory and observation: for instance, fields went from being described as "immaterial" to being described as "material." But despite this verbal realignment, the fact remains that Cartesian materialism had been falsified. We cannot make do with the exiguous range of entities and types of causation favored by Descartes: extended bodies and contact causation.

The process of falsification and expansion only continued with later discoveries. More and more types of physical reality had to be recognized, such as radio waves and X-rays, and the sheer versatility of matter became increasingly evident. The concept of energy added to this diversification, as many species of energy were admitted—kinetic energy, chemical energy, gravitational energy, electro-magnetic energy, nuclear energy, and so on. Forces multiplied, as did particle types. Indeed, the question whether all these entities could be classed univocally together as "matter" became a real issue: might the term not be equivocal or at best a family resemblance term? The unity of matter/energy starts to look shaky when entities of such disparate kinds are admitted. Maybe the term "matter" had outlived its usefulness—and it was never very precise to begin with (is it really any better than "stuff" or "substance"?). In any case, diversity was the order of the day—heterogeneity, multiplicity of forms.

Let me just indicate by list the kind of ontological plurality that has come to characterize physics (and the univocity of *this* term also must come into question). There are electrons and protons, which differ considerably in respect of mass, and which enjoy different electric charge, as well as occupying different parts of the atom (shell and nucleus, respectively). It is a real issue whether these can both be said to be composed of the same stuff, and even if they can they differ enormously in their "form." We have charged particles of two types, positive and negative, which have widely different attractive and repulsive properties, as well as particles that have no charge at all. It must be very different forms of matter that can underlie such differences of behavior: differently charged bits of matter must reflect a deep ontological division. Most matter is said to possess mass (inertia), but it is often noted that neutrinos have no mass, and it is not clear that fields can be said to have mass in any recognizable sense (the concept of mass itself seems to resolve itself into a motley of distinct subspecies). The distinction between matter and fields, if it is taken to be fundamental, reveals another deep fissure in the concept of the "physical": certainly a counterfactual understanding of the notion of a field shows it to be quite different from "substantial" matter. The four basic forces—gravitational, electromagnetic, strong, and weak—look fundamentally different from each other, despite attempts at unification: this is nothing like the Cartesian picture of a single contact force borne by bodies in motion.

Light has always seemed to stand apart from more sluggish forms of matter and has not always been so designated. The constancy of its speed is a mark of its distinction from other physical things, and the fact that this speed cannot be exceeded sets it totally apart: there are those things that can (and must) travel at c and those things that cannot—and this is a basic divide in nature. Particles and waves are ontologically quite distinct, being different forms in which nature can be configured; and one can understand the temptation to deny that waves propagated in empty space deserve to be called material (however invidious this may be). Nowadays we hear of dark matter, which has mysterious repulsive properties and which apparently is quite different from the milkier kind. Ether was once regarded as a very thin and light form of matter, quite different from the chunkier chemical elements—yet still matter of some sort. If there had been such a thing, it would have been matter in a radically different form. All around we see plurality, division, and a sense of strain in the traditional fixed notion of a universal material substance. We are confronted by irreducible variety.

Given all this, it would not be surprising to find historical resistance to the introduction of new categories into physics. Something is taken as a paradigm of the material—say, solid contacting macroscopic bodies—and then other supposed entities seem not to live up to its exacting standards, and thus are denied the honorific label "matter." Fields don't fit; light doesn't either; even elementary particles scarcely qualify. Much semantic twisting and turning ensues, and heated debates, and verbal stipulations. It is decided that we will need a new general term to accommodate the new borderline entities—in which case, we may as well re-define "matter" by means of the new term. The right lesson surely is that there is no paradigm of the material: there are simply many forms that so-called "matter" can take, and these can be irreducibly various. (Or perhaps better, in view of more recent formulations of physics, energy can take many forms.) It would then be natural to go to the opposite extreme and assert that there is nothing unitary behind these different forms, so that we should simply stop classifying them together. Even the idea of a unitary science called "physics" should be abandoned. But I think there are reasons for not going quite that far—which I will get to later. What I want to stress now is the great diversity of material forms (compare the diversity of life forms—but does that mean the concept of "life" should be abandoned?). Even basic entities of physics, not built of others, are quite different ontologically from each other. Cartesian uniformity has turned out not to hold of nature. Cartesian materialism is not true of physics, let alone the mind.[1]

---

[1] Descartes oversimplified the natural world in his philosophy of physics—he underestimated its complexity and variety (and strangeness). He is (ironically) a kind of procrustean reductionist about the "physical." A not dissimilar oversimplification or reductionism afflicts thinking about the mind in Descartes' era (and later), in the form of the assumption that everything mental is an "idea"—either a sense impression or a faint copy of a sense impression. Everything physical is an extended body in space, and everything mental is a perception in consciousness—a discrete and

These reflections leave open a hypothesis that is worth considering: *that consciousness itself is yet another form of matter.* Just as there is electrically neutral matter and electrically charged matter, so there is insentient matter and sentient matter—these are just two *types* of matter. An electromagnetic field is a type of material reality, and so is consciousness. Alternatively, consciousness is one form of energy, along with kinetic energy or electrical energy. If this hypothesis is true, then consciousness is material after all—though not in the Cartesian sense. The general conception I am working with is that matter/energy is the underlying substance of the universe, and it may ultimately be unitary, but it can take widely different forms—with consciousness as just one of them. That is the abstract metaphysical picture: an underlying unity (possibly quite removed from our current conceptions) combined with startlingly different modes of that unity. Matter/energy is thus conceived as indefinitely plastic and versatile, as witness physics as we have it, and among its many manifestations is consciousness.[2] Consciousness is one of the modes that matter can appear in—one mode of its being. Call the basic stuff of the world X and suppose X to be uniform

homogeneous mental content. Locke, Berkeley, and Hume favored the sensory as their basic type of the mental, while Descartes assimilated everything to "thoughts." There is a drive in all three philosophers to find a uniformity in the realm of the mental analogous to that claimed for the physical: extended bodies here, conscious ideas or thoughts there. But that psychological monism is as suspect as the analogous physical monism, as later developments demonstrated. Sensations and thoughts are not the same; will is not the same as perception; intention differs from desire; emotions have their own character; images are not percepts; not everything mental is a content of consciousness; maybe some mental items are not even conscious. The so-called "mental" is itself heterogeneous and multi-faceted, with a complexity and variety not recognized by the earlier thinkers. And a very similar story holds for language: not every word is a name of something; not all sentences are indicative; speech acts are various; quantifiers are semantically distinct from singular terms; not all contexts are extensional; not every utterance is descriptive. Again, variety triumphs over uniformity: earlier thinkers had oversimplified, underestimating how different the things we call "language" can be. There is thus no "single essence" of language, or mind, or matter. What we call "mental" is a mixed bag, what we call "language" is a motley crew, and what we call "physical" is a hodge-podge. In all three areas, the movement has been towards plurality, away from paradigms, and antiprocrustean in tendency. I suspect that many contemporary philosophers would immediately accept the plurality point about mind and language, but are still stuck with Cartesian homogeneity when it comes to the "physical." (Wittgenstein is said to have wanted to use as a motto the line from Shakespeare's *King Lear*: "Come, sir, I'll teach you differences," said by Kent to Oswald, in Act 1, Scene 4. Echoing this sentiment, in the same scene, the Fool asks Lear: "Dost know the difference, my boy, between a bitter fool and a sweet fool?" Lear laconically replies, "No, lad. Teach me." It seems that in our callow days we presume uniformity and only with the coming of maturity do we accept variety.)

[2] We could think of matter/energy as a determinable, with consciousness one of its many determinates. But we could also allow for disunity all the way down, so that there is no single natural kind denoted by "matter/energy." Then the hypothesis would be that consciousness is no more "nonmaterial" than the other forms of nature traditionally dubbed "material." One view might be that "matter/energy" is a family resemblance concept (as has been maintained for "mind" and "language"), with no single essence. There are several options regarding the semantics of "matter/energy" aside from the one I favor in the text, which might suit those with different predilections from mine. The essential point is to resist the kind of stark dualism promoted by Descartes, which rests upon his narrow and outdated conception of the "physical."

(knowably or not): then consciousness is a mode of X, like every other natural phenomenon. Putting it this way we can avoid the heavy historical connotations of the word "matter," with all its biases and baggage: "X" is just a name for what composes everything natural. The important point is that consciousness is a form of the same thing that everything else is a form of: electrons, protons, fields, neurons, elephants, stars—whether we call this thing "matter" or not. That, at any rate, is a hypothesis that reflection on the progress of physics suggests for consideration. It is a hypothesis that arises from the liberalization of the concept of matter from Cartesian restrictions. We must now ask if there is any reason to favor it and any reason to reject it.

Three (nonapodictic) reasons might be offered in support of the hypothesis. First, it is hard to see what *else* consciousness could be. Consciousness exists *in* a world of matter/energy, not outside of that world (as God and his angels might be supposed to exist), and depends essentially on (other) forms of matter, causally and otherwise. Given that Cartesian dualism has daunting problems, and is not even clearly intelligible, there doesn't seem much of an alternative to supposing mind to be in *some* way a modification of matter: the question is, in *what* way. What we really want to know, in thinking about the mind-body problem, is how it is *possible* for consciousness to be what we know that it *must* be. What *kind* of materialism (if we must use the term) is defensible? Put differently, consciousness must be an aspect of the same world that (other) forms of matter are also aspects of, notably the brain. Organisms are modes of matter, with some distinctive properties, and consciousness is a biological property of organisms; so it is only natural to assume that consciousness too is a form of matter. To say that it is a form of an "immaterial" substance is to fly in the face of the obvious truth that consciousness is part of the world of embodied organisms—not a separate parallel world, with strange causal connections to the regular corporeal world. There is really nothing else for consciousness to be a mode *of* than the very stuff that everything else is a mode of.[3]

Secondly, conservation laws in physics preclude the idea of a radically new kind of stuff, energy or matter, coming into existence. So when consciousness came to exist, no new substance was added to the world: old stuff simply took on a new form. Descartes' dualism violates conservation, since extra causal powers—extra energy—are introduced by the injection of mind into the world. His immaterial substance is an independent source of energy and hence motion, so that conservation is bluntly

---

[3] A pressing question about so-called "immaterial" substance is why it is not just another kind of "material" substance—why the deep dichotomy? The reason, historically, is that Descartes took all material substance to consist by definition of extended bodies interacting by contact, and he couldn't see how mind could be extended and in contact—*hence* mind is not material. But if we abandon his definition of matter, as we have to anyway, what stands in the way of saying that the substance of the mind is just material substance—the nonextended noncontacting kind? We need some definite conception of the material in order to be able to locate mind outside the realm of the material—and that is precisely what we don't have. Instead of a dualism of the material and the immaterial, why not a dualism (or pluralism) *within* the material (or "material," to add the requisite scare-quotes)?

violated. A better view is that pre-existing matter takes new forms in cosmic history—from galaxies to organisms—and consciousness must itself be a form of what existed earlier. But there must be a fundamental constancy in the underlying substance of the world, whatever that may be: so consciousness must be a variant on this substance, not a new type of substance.

Energy is plausibly the fundamental conserved substance, so consciousness has to be a form of energy—a form of the very same thing that electricity and mass are forms of. Moreover, energy in its various forms can be transformed into mental energy—as when the chemical energy in food (deriving from solar energy) is converted into causally efficacious acts of will and other mental work. The energy that powers the mind is nothing other than the energy that exists in various physical forms; and so it is plausible to suppose that the mind itself is a manifestation of that energy. Certainly, the electromagnetic energy of the brain has everything to do with the energy exhibited by the mind, i.e., its ability to do work. The world we observe is a world where conservation is the norm—where nothing fundamentally new comes to be. Novelty comes from recombination, not from new basic realities. Similarly, when consciousness fades away, nothing basic goes out of existence; rather, it changes form—going back to material forms of other kinds. Such continuity suggests that consciousness is just matter/energy in one of its many guises. Kinds of matter/energy can go out of existence, or be created, but the underlying stuff stays constant—as when particles change into other kinds of particles or electrical energy converts to kinetic energy.[4] And the same is true of the kind of matter/energy we call consciousness. Consciousness is a temporary form that universal matter/energy has taken, along with other forms.[5]

---

[4] Heisenberg says: "The experiments have at the same time shown that the particles can be created from other particles or simply from the kinetic energy of such particles, and they can again disintegrate into other particles. Actually the experiments have shown the complete mutability of matter. All the elementary particles can, at sufficiently high energies, be transmuted into other particles, or they can simply be created from kinetic energy and can be annihilated into energy, for instance into radiation. Therefore, we have here actually the final proof for the unity of matter. All the elementary particles are made of the same substance, which we may call energy or universal matter; they are just different forms in which matter can appear": *Physics and Philosophy* (HarperPerennial: New York, 1958), 133–34. It was actually this passage that gave me the idea for the present paper, and for its title: for if all forms of matter are variants on the same fundamental substance, heterogeneous as they are, why not suppose consciousness itself to be a form of this very substance? Heisenberg takes transmutability experiments to be strong evidence for the unity of the basic stuff, and consciousness is just another transmutation this stuff can undergo—as suggested by energy conservation and the origin of everything in the material of the big bang. Since the same stuff can take radically different forms in the realm of physics, why not add yet another form to its list of accomplishments?

[5] Of course, there is every reason to suppose that consciousness will be entirely obliterated at some point in the distant future—if not by the death of the sun, then by the big crunch that follows cyclically on the big bang. When it does disappear the universe will not lose any of its diabolical energy, so that consciousness cannot be an original source of energy. As the dinosaurs were a form of matter that is now extinct, with no loss of the universe's total energy, so consciousness will one day become extinct, and the universe will lose nothing of its surging energetic power. The energy contained in consciousness will simply be re-absorbed by the universe, as it mindlessly seethes and hurtles. The forms come and go, but the universal substance lives forever.

Third, and connected, the big bang contained all the materials for generating the universe from then on. New particles came to be in the first few moments, and new forces too, but everything had to be implicit in the initial super-hot plasma: everything that followed had to be a form of what was there at the start. All novelty works with the raw materials of the primal singularity (just as the big bang itself had to be a conservation of what was there earlier). But if so, then consciousness must be somehow implicit in the big bang too: it must be a working out of the matter/energy there at that instant, like planets and organisms. Some of what is new is mere recombination (this is probably true of organisms, now that the *élan vital* has lost its appeal), and some of it consists in new forms of what was there earlier (were gravitation and electric charge explicitly present at the very earliest stages?). But nothing we see in the universe now belongs to neither category: so consciousness must be one or the other—and the recombination view is not credible. Consciousness must be a new form of the stuff that was present in the first moments of the universe—one of the modes of which matter is capable. There was no ghostly parallel big bang in which immaterial stuff was minted, with consciousness a form of that: consciousness had to have its origin in the very event that originated all the matter of the universe. If we think of the history of the universe since the first moments of the big bang as a process of differentiation in matter, then consciousness is one of the many ways in which matter came to be differentiated—and clearly it specializes in such differentiation. Matter is nothing if not protean.[6]

I could put the point this way: granted that matter/energy is demonstrably versatile, taking many forms, why *not* suppose that consciousness is one of its many protean achievements? Once the concept is cut free of its Cartesian restrictions, it appears able to cover a wide range of natural phenomena, often of very different natures (e.g., fields and particles, as well life forms). With the mechanistic theory of matter abandoned, what reason is there to *deny* that consciousness is matter in one its more exotic manifestations? That is to say, consciousness is one form that the fundamental stuff of the world can assume (which for historical reasons we persist in calling "matter"). Or to put it conditionally: *if* there is a "universal matter," a single unitary world-substance, then why *not* say that consciousness is a species of it—

---

[6] Conceivably, the material of the big bang might not have been such as to be capable of sustaining the variety of forms that we see in the universe today. It might not have had the power to form galaxies, and organisms, and conscious minds. It might have been capable of nothing more than creating a vast sea of formless gas. It is just lucky that *our* matter has the resources it has shown itself to have—so adaptable, so versatile. Fortunately for us, our actual matter has the ability to form conscious minds! Note, too, that it can take a long time for matter to discover its full potential—some 14 billion years before conscious organisms appeared (unless they have done so elsewhere at a much earlier date). In another 14 billion years who is to say what other innovations it might initiate? What other brand new forms might it contrive? Will they be as impressive as life and consciousness? Can we even conceive of what they might be? It is, after all, still quite early in the creative career of matter/energy. The stuff is evidently a pretty potent brew, judging by its track record: it might have all sorts of ideas up its sleeve for the future.

along with electrons, protons, fields, organisms, and so on? Only Descartes' restrictive mechanistic view of matter could ground an insistence on substance dualism: but that lapsed long ago, leaving us with a happier monism (combined with pluralism). In fact, we are left with a reversion to something like Aristotle's outlook: that the same basic stuff can take many forms—from rocks, to plants, to animals, and to conscious subjects. There is a pluralism of forms, if you like, combined with a monism of substance. What physics has taught us is that even in the world of inanimate things, even indeed in the subchemical world, matter is a motley collection—with a variety of forms all the way down. Then what principled reason exists to deny that consciousness is one of these varieties?[7]

The view I am describing—it has no established name that I am aware of—may strike some readers as but a variant on views with which they are already familiar. But I don't think it is the same as any of the usual positions, and I must now explain why not. It is not a form of panpsychism—indeed, it is in some ways the opposite metaphysical position. Panpsychism in effect holds that matter is one form of consciousness, since the indisputably mental nature of consciousness is held to extend to areas generally considered not mental in nature (elementary particles, even space). I am, to the contrary, defending a view that says that the indisputably material nature of, say, electricity or elementary particles, extends to an area not generally considered material, viz. the mind. The view might aptly be labeled *panmaterialism* (but see below on whether the view deserves to be called a kind of "materialism" in the current sense of that slippery term). It is true that panpsychism maintains that matter and mind are not incompatible, being found locked together in all the same entities, right down to elementary particles. But there is no commitment in my view to the idea that mind extends any further than its conventional boundaries. Neither am I saying merely that what is mental also has material aspects; I am saying that mental aspects *are* material aspects—that consciousness *is* matter in one of its many manifestations (as the electromagnetic field *is* a form of matter/energy or as a proton *is* one form of matter). I am saying that consciousness is material *as such*, not by way of association with other things already so classified.[8]

---

[7] Compare denying that thoughts are mental because they lack the qualitative character of sensations, or denying that commands are sentences because they are not assertions, or denying that bacteria are organisms because they are not visible. These are all instances of what might be called "the fallacy of the misplaced paradigm": arbitrarily selecting a paradigm, perhaps for boring historical reasons, and then insisting that certain things don't fall into a particular category because they fail to conform to that paradigm. My point is that denying that consciousness is a form of matter is an instance of that fallacy: claiming that just because consciousness doesn't resemble solid bodies in space interacting by contact causation it isn't "material" (as gravitational attraction and electromagnetism were once deemed not "material").

[8] Superficially, I am saying just what Galen Strawson is saying when he declares experiences "physical" just as such, without regard to their reducibility to something already deemed physical (such as neural states): see his "Real Materialism" in *Real Materialism and Other Essays* (Oxford University Press: Oxford, 2008). But beneath the surface our views are actually rather different. I

This formulation may now prompt the (relieved) suspicion that I am simply peddling the old identity theory of mind and brain with some irrelevant and obscure ideas drawn from fancy physics. After all, I am saying that conscious states *are* states of matter—material states: and couldn't we rephrase the traditional identity theory by saying that conscious states are forms that matter can take (neurons firing and so on)? But this would be to misunderstand the present view grossly. I am *not* saying that we can form an informative identity statement using a mental term and a term for a physical state, as in "pain is C-fiber firing"; I am saying that conscious states are themselves *already* forms of matter, *without having to form such an identity statement*. Being a pain, say, is, considered in itself, a modification of matter, without the need for any reduction to something antecedently recognized to be material. Indeed, I would deny that the usual kinds of identity statements are true, because the form of matter that consciousness constitutes is *not* the same as the form that the brain constitutes (as we conceive it). When matter/energy forms electrons and protons it takes particular forms, and these are the ultimate elements of the brain; but when matter forms consciousness it is not thereby forming electrons and protons. Consciousness is a form of matter in its own right, whatever its relation to other forms of matter might be. That at least is the view, and it is abundantly clear that it is not a variant of the old identity theory.

It is, if you insist, a new kind of identity theory, since it holds that conscious states are identical to matter of a certain sort (the conscious sort): but there is no way to specify what sort except by invoking consciousness itself.[9] Compare: the electromagnetic field is a type of matter/energy, but we do not try to say what type by using other terms drawn from physics—any more than by saying that electrons are a type of matter we are committed to specifying them in other terms (as it might be, terms for protons!). All these things are *as such* forms of matter: they need no translation or reduction to grant them that status. Consciousness is matter (= the unitary world-substance) that has taken one form after being of another form—all the way back to the big bang. Compare: galaxies are one form that

---

base my claim on the versatility of matter, while he does not. I find it necessary to link consciousness to other forms of matter in order to motivate my claim, while he does not. I am much more leery of the terms "physicalism" and "materialism" than he is. He thinks the experiential is of the essence of matter in general (panpsychism), while I do not. Still, our views have certain affinities, particularly with regard to our attitude towards the distorting influence of old-style Cartesian materialism.

[9] Thus we can assert the proposition "The mental state of pain is identical to a material state" and mean it quite literally. But if we are asked what material state it is that we hold pain to be identical with, the only answer we can provide is "The material state of being in pain." For we hold that pain is a material state in and of itself, independently of whether it can be specified in other terms. This is really nothing like classic identity theories. Compare classifying an electromagnetic field as a material state of space. If we are asked what that material state may be, we can only reply "The material state of containing an electromagnetic field." This will clearly not satisfy a diehard mechanist who wants a reduction of electromagnetic fields to mechanical interactions of solid particles. But then why should we seek to satisfy such a mechanist? He simply wields an entirely optional conception of the material, which we are under no obligation to accept.

matter has taken after having existed in a gaseous form. Matter is not static: it keeps reshaping itself into novel incarnations. There is a kind of restlessness to it—an appetite for experimentation. It can reinvent itself as consciousness without having to do so via a reduction to brain states of the usual kind.

So I must be some sort of dualist! For I hold that consciousness is one form of matter, while allowing that there are other distinct forms too—and that there are causal and other connections between these contrasting categories of matter. I am a "form dualist," it may be said, despite my avowed substance monism. That is not so far off, except that I am actually a *pluralist* of forms. What I am not is a dualist or pluralist of basic substances, since I hold that all reality is (most probably) made up of forms of a single basic stuff—stuff that is, in a certain sense, highly abstract and removed from sensory observation. We can refer to this stuff as "matter," remembering that this word and its cognates can be correctly used in such phrases as "subject-matter," "what's the matter?" and "material witness"—that is, it has no fixed connotation tying it to Cartesian materialism. I am a pluralist of forms because I hold that matter takes *many* forms, not just two: matter is far more adaptable and versatile than the old Cartesian dualism allowed. There are positively charged particles, negatively charged particles, and neutral particles (protons, electrons, and neutrons, respectively)—this is pluralism about the elementary forms of matter. I add consciousness as yet another form of matter, along with myriad other things. This is a far cry from traditional Cartesian dualism, and is best seen as a repudiation of that kind of metaphysical picture. In sum: I don't think consciousness is identical to a state of the brain (specified in neural terms), but I advocate regarding it as material anyway—in virtue of its own nature. I am, like Aristotle, a monist about substance but a pluralist about its forms—which is a very different metaphysical architectonic from that of standard Cartesian dualism.[10]

Neither is the view a form of emergence, holding that conscious states earn the name of "material" by virtue of arising from states already declared material, such as neural states. Maybe conscious states do so emerge, but the view does not require that to be so. They are forms of matter because (roughly) everything is, not because they stand in an emergence relation to something already classified as material. Perhaps some forms of matter can emerge from other forms (without genuine reduction), in which case the form of matter called "consciousness" might emerge from some other form of matter: but the doctrine I am defending is neutral on that

---

[10] Descartes recognizes two fundamental substances, whose essences are extension and thought. I prefer a single substance, with one essence, and neutral between mind and the rest of nature. I think it is wrong to define matter as extension, and mind as thought—too restrictive in both cases. I think there are indefinitely many forms of natural things, each with its own essence, with no fundamental duality running through them. As I understand him, my view is similar to Spinoza's view (but without God). The whole dichotomy of "mental" and "physical" is deeply flawed as a metaphysical delineation (though the words can still be useful in specific contexts). The traditional idea that reality divides into the "mental," the "physical," and the "abstract" seems to me beyond redemption and should be scrapped entirely.

question—it holds that consciousness is already as such is a form of matter. So if it turns out that is there no such emergence, then consciousness *still* counts as a form of matter, on the present view. The best label I can come up with for the view I am defending is "boring materialism," but the label is unsatisfactory for at least two reasons. The first is that the word "materialism" has already been done to death as the name of a different position, and should only be used with the utmost caution anyway.[11] The second is that the view is not actually all that boring: it reflects some substantial claims about the nature of the material (sorry!) world. It is really quite interesting—is it not?—to conclude that consciousness is on a par with other more orthodox material things, such as electromagnetic fields and bodies in motion. Perhaps the view might be (tendentiously) named "straightforward materialism" or "conservative materialism" or even "uncontroversial materialism." In any case, it holds that there is something true and illuminating about describing consciousness as a form of matter, however we name the view.[12]

I just said that, with the proper caveats entered, we could describe the view defended here as a type of "materialism." But I want to emphasize in the strongest terms that such a label is potentially massively misleading. In a certain very thin sense the label is linguistically justifiable, since the view does indeed maintain that consciousness is a form of matter, where "matter" is the name of the fundamental world stuff (what I earlier called "X"). But it is remote from any doctrine usually so called, in that it does not claim that we can reduce consciousness to something specifiable in other terms—still less that the terms of physics, as we have it, provide a sufficient conceptual basis for understanding consciousness. Furthermore, the view is spiritually the opposite of traditional materialism, since it resists the kind of ontological leveling recommended by such doctrines: it insists on the irreducible variety of the forms of matter, not their inter-translatability (i.e., the *dis*unity of physics). The claim that mind can be reduced to the properties allowed by Cartesian materialism *does* count as a real "thick" version of materialism, since it supplies a

---

[11] Here I side with Chomsky: see his entry in *A Companion to Philosophy of Mind* (Basil Blackwell: Oxford, 1994), ed. Samuel Guttenplan. However, I still think the term "material" can play a useful role in philosophy of mind, though it needs to be suitably qualified and explained.

[12] The following is a far-flung analogy intended to illustrate the semantic situation: consciousness is a form of matter in the same kind of way that democracy is a form of tyranny. The latter statement looks paradoxical at first, because we are fixating on certain forms of tyranny (monarchy, aristocracy, the police state). But then it is pointed out that in a democracy the majority can dictate to the minority, overruling their interests—hence "the tyranny of the majority." Now our perspective is changed, in such a way that we come to see democracy in a new light—an unobvious feature of democracy turns out to unite it with more obvious forms of tyranny. Classifying (simple) democracy with other forms of tyranny thus has a point to it—it is not merely a linguistic stipulation. Similarly, calling consciousness a "form of matter" initially suggests certain paradigms of matter, from which consciousness seems admittedly remote. But then certain considerations are adduced to dislodge the hold of these paradigms, and to include consciousness in the extension of "material"; thereupon the classification comes to seem quite reasonable. (Reader: please don't read more into this analogy than I intend.)

uniform framework of material properties with which to account for the world. But on the more expansive view of physics, in which irreducible heterogeneity obtains, there is no attempt to squeeze the varieties of matter/energy into a single conceptual framework. Either we regard "matter" as something like a family-resemblance term, with both consciousness and electrons as part of its mixed extension; or we adopt the metaphysical view I prefer, which holds that there is a single world-substance, remote from observation, highly versatile, which can take a myriad of distinct forms. In neither case do we have a view that resembles traditional materialism, i.e., the kind founded on the Cartesian picture of (nonmental) nature.

Indeed, I can sympathize with someone who regards the view here defended as a type of mentalism or idealism, since it holds that one form of matter at least consists of consciousness—and the other forms are variants on what consciousness itself is a variant on. I am saying, after all, that the so-called material world is, at least in part, irreducibly conscious. But the labels don't matter much: the point is that what we now call "matter" has come to be far removed from the earlier mechanistic picture—so far removed that it no longer seems reasonable to deny that consciousness can also come within its scope. Earlier forms of matter have given rise to the form we call consciousness—with no addition of a new type of substance—and so we can regard consciousness as just the latest in the achievements of the stuff that was there all along, and which goes back at least as far as the big bang. That is the essential core of the position.[13]

There are, of course, correlations between consciousness and other forms of matter, specifically electrical activity in the brain (mediated by neuronal activity). These may even be described as lawlike, in some relatively weak sense. It may also be that there is a supervenience relation between consciousness and other material forms. Causal relations between electrical events in the brain and conscious events are evident. But these kinds of interrelations between forms of matter are common-

---

[13] Although my type of "materialism" doesn't depend on any reductionism or even emergence claim with regard to brain states, much of my motivation for adopting the view relies on discerning *links* to other material things—the mind's immanence in the biological world, the sharing of energy with physical systems, the origin of everything natural in the big bang. A possible world in which all these links were abrogated (if such a world there could be) would not for me be a world in which consciousness is plausibly described as a form of matter; in that world it might be a form of some different natural kind (as it might be, ectoplasm). So there is a sense in which I am counting consciousness as "material" because of its *connections* to things already classified as "material"—which function as paradigms of some sort. Without something like this to anchor the word it seems entirely empty. Still, my type of "materialism" is a far cry from traditional doctrines so named, since they *identify* consciousness with something already deemed "material," or at least view it as emerging from such. I suspect that the earlier enlargements of the extension of "material" occasioned by gravity, electromagnetic fields, and energy traded upon the same kinds of links: these things came to called "material" because of their connections to entities already deemed so—massive bodies moving in space, specifically. I doubt that anyone would have been willing to designate electromagnetic fields as "material" if no connections to other "material" things had existed. The enlargement doesn't proceed from direct inspection of the disputed entity, but from noting its connections to things not in dispute. So it should be, I contend, with consciousness.

place in other areas. Consider an electromagnetic field and the force exerted on a body at some point of space within that field: the latter results from the former (the electromagnetic energy at that location, which has amplitude and direction), and there may well be a supervenience relation here (the amount and direction of force exerted is supervenient on the field properties). But this should not lead us to conflate force and field: the field exists whether or not it is exerting any force on a body, and the very same force could be exerted by a different agency (say, brute contact causation). Supervenience is in general a very weak relation and does not lead to any form of reduction. Indeed, it is compatible with both terms of the relation having a common cause, since it is simply necessary covariation. It is thus logically possible that both electrical-neuronal activity and conscious occurrences have a common cause, perhaps in some deeper property of matter. When a charged particle moves it generates a magnetic field, whose properties are a function of the charge and the motion: but this should not lead us to assert that magnetism is nothing but electricity plus motion. Similarly, consciousness may covary with other forms of matter, such as electricity in the brain (in the form of neural impulses), but that is not sufficient to inspire the thesis that consciousness just is that other form.

Nor is it in virtue of such interrelations that we are entitled to speak of consciousness as itself a form of matter. I think it is significant that energy may be transformed from electrical to mental, so that the ability of the mind to do work results from energy that existed in another antecedent form: but we should not call the mind material simply because of such a transformation—any more than the fact that kinetic energy can be transformed to electrical energy is the *reason* we call electrical energy a form of matter. Consciousness is indeed continuous with the others forms of matter/energy, meshing with other forms, sharing their energy content, causally interacting with them—and this is an important part of the reason to suppose that there is a single reality underlying both consciousness and the rest of nature. But that is not a *definitional* truth—it is not *why* consciousness merits the label "form of matter." So if my position here is aptly termed "materialism" it is not for the usual sorts of reasons. In fact, I would prefer to shun the label as being misleading in the contemporary climate and too closely tied to discredited ideas of matter—though I am prepared to accept it as literally quite correct. We could, I suppose, call the position "multiple variants of a single reality-ism," grotesque as that may be. Or I might even settle for "antimaterialist materialism," relishing the oxymoron.[14]

I have said nothing substantive about what consciousness is, leaving the concept at a very intuitive level. Different views about that question might be dovetailed into the position I defend in this paper. If we regard consciousness as consisting in states

---

[14] One does wonder how much harm is done to philosophy by the hunger to find a brief and catchy label for what may be a complex and subtle position. Misplaced assimilations and fruitless sloganeering are the inevitable outcome. I leave it to the reader to think of examples ("realism" and "cognitivism" in ethics spring to my mind; or "the linguistic turn"). But it is hard to do without them.

there is something it's like to have, then the position is that what-it's-like states are forms of matter. If consciousness is defined as that which is accessible to introspection, then there is a form of matter that is an object of introspection. If we define consciousness in terms of intentionality, then some forms of matter have intentionality. I myself favor the view that consciousness is a type of knowing; on that view, then, there exists a knowing form of matter. This is all to say that among the many and varied achievements of matter/energy—being positively or negatively charged, neutral, massy and gravitational, field-like, particle-like, wavelike, dark, visible, bounded and unbounded, ridiculously dense, impossibly light, super-fast, ploddingly slow, easily transformed, constant in total quantity, possibly indeterminate and probabilistic, nonlocally entangled, maybe multiple world-y—among all these forms there is a form that consists in being subjective or introspectible or intentional or epistemic.[15] If matter can do all those varied things, can't it also contrive to be conscious? Isn't this really a *datum* of the world, once traditional prejudices are

---

[15] I hope that by now the temptation to argue that consciousness cannot fall into the category of matter, because matter cannot be subjective or intentional or epistemic, has lapsed. That style of arguing presupposes that nothing can count as matter (or a form of it) unless it conforms to certain paradigms, and then claiming that consciousness does not match those paradigms. But matter can have a potentially unlimited variety of forms, among which there is immense variation, and it is arbitrary to select some forms as setting the standard above all others. What if the earliest solid results of science, mathematically formulated, had concerned consciousness, and not other phenomena? And what if we had called the stuff of consciousness "matter" back then (I mean this purely orthographically)? Then the thing we called "matter" (i.e., the stuff of consciousness) would have the prestige of all successful science. Other departments of learning lagged behind, with even simple mechanics proving intractable. There would be an understandable reluctance to extend the word "matter" beyond the case of consciousness, even though "matter" is defined as covering everything that is a form of what consciousness is a form of—because of its honorific connotations. Well, it is just like that with our word "matter" (or "physical") as now used: the solid results came with phenomena not including consciousness, with the study of consciousness lagging behind, so there is a reluctance to extend the honorific term to consciousness. But consciousness *must* be an achievement of matter, for the reasons cited in the text (and others)—so we need to get over our reluctance. Nor is it a good objection to this that physics is "objective" while the study of consciousness is "subjective." The answer to that is that physics (the general study of all matter) is *not* (wholly) objective, given that it includes consciousness, and consciousness is subjective (or requires a subjective mode of understanding). (I am alluding here to Thomas Nagel's *The View From Nowhere* (Oxford University Press: New York, 1986)). Physics is partly objective and partly subjective, *if* it includes all the forms that matter can assume. It is true that physics, as we have it now, does not include consciousness, and hence is wholly objective; but that is for boring institutional and historical reasons. There was a time when physics didn't include force fields—that doesn't mean that force fields are not a form of matter. Someone might then have argued that physics is the science of the *palpable* and fields are not palpable, so they are not part of physics. It is similarly question begging to argue that consciousness cannot be a form of matter because all forms of matter are objective and consciousness is subjective. If consciousness *is* a form of matter, then physics (the general theory of matter) is not completely objective. (I am not suggesting that Nagel himself argues in the style I am opposing, but I can well imagine someone using his ideas to argue that way.) The right thing to say, from a wider perspective, is that one part of physics is not reducible to another part—but that we already knew before we ever came to consider consciousness. Subjective physics is not reducible to objective physics (as electricity physics is not reducible to gravity physics). Adding consciousness to physics is just adding to the disunity of physics we have already had to recognize.

shed? Or perhaps I should say: why deny that consciousness is matter in action, given all the other remarkable things that matter can do? Whatever we take to be the essence of consciousness, matter can have that essence—as it can have all the other kinds of essence we see in the natural world. For what *else* could have that essence?

Does this solve "the problem of consciousness"? Does it make the "hard problem" soft, and safe? Does it remove the "mystery"? None of the above: for two reasons. One is that we still don't know how the brain generates consciousness—we haven't solved the problem of emergence. We now know that consciousness is one form that matter can take, but we don't know how this form can arise from the forms present in the brain—which consciousness appears to do. The neurons are the causal basis of consciousness, and both are forms of matter, but we have no idea how the latter form arises from the former. Nor do we know how in cosmic history pre-conscious forms of matter could have produced the conscious forms (as we know how some particles were produced by others during the big bang or how the chemical elements were formed in the interior of stars by atomic fusion). If we opt for the view that the emergence of consciousness from atoms is merely a brute fact, a primitive feat of nature, then we are saying that one form of matter produces another form of a totally different sort—a logically possible position but surely an unappealing one. So agreeing that consciousness is a form of matter leaves the mind-body problem pretty much where it was.

What we have learned is rather that Cartesian materialism has biased us away from acknowledging that consciousness can be a type of matter in its own right, and that a position of pluralism about the varieties of matter is preferable (combined perhaps with an underlying monism). The mind-body problem is now about how one form of matter can be appropriately connected with other forms—where this is no longer a question about whether something called "materialism" is true. The problem is no easier than before; it is just better formulated. Perhaps, though, the pressure towards eliminativism has been relieved somewhat, in so far as that pressure depended on accepting Cartesian materialism. In the early days of astronomy and physics, there was a similar eliminativism in the air as new discoveries led to ontological expansion: some denied that Newton's gravity could exist, since it involved occult action-at-a-distance; the electrical fields of Faraday and Maxwell were regarded with suspicion for a long while; even now the situation in quantum mechanics feeds antirealist sentiment. But if we find ourselves having to admit these strange new entities, then is it so hard to accept that consciousness—hardly a recent discovery—should also be admitted, despite its remoteness from the Cartesian paradigm of the material? After all, as we have seen, there are lots of real things—in physics, no less—that fail to conform to the old ideal of a bounded particle suspended in empty space and acting by means of contact causation. Calling consciousness "material" affords it shelter from the rampaging eliminativists of the world.

The second reason that the view here defended does not make the mysteries disappear is that merely to call something a form of matter—and for this to be simply and literally true—does *not* mean it is devoid of mystery. Philosophers of a certain stripe have always fondly supposed that if it's material it can't be mysterious, so they sought to ensure that the mystery of consciousness would dissolve by its assimilation to the material. But this is a very naïve view of physics, which is in fact rife with mystery. I can't discuss this topic fully now (see the rest of this book) but it is widely agreed that most of the advances in empirical physics have just opened up new areas of mystery. The most famous example is Newton's theory of gravitation: this provided a marvelous new general theory of motion, but the postulation of gravity was widely felt to introduce a mystery into the heart of the theory. Less celebrated, but equally potent, is the Faraday-Maxwell theory of electricity and magnetism, also discussed earlier: we likewise have mysterious action-at-a-distance, but we also have the mystery of what an electric charge *is*, of how it generates a field, and of what a field is and how it exerts force on an affected body. Quantum physics of course is nothing *but* mysteries—with a bit of useful mathematics to take the pain away.

Physics raises many deep ontological questions that are not answered by its own empirical and mathematical methods. So deciding that it is good manners to admit consciousness to the Matter Club [16] is no prelude to declaring it mystery-free. Indeed, granted that physics itself is full of mysteries, recognition of club membership only makes it more intelligible that consciousness should be the mystery it palpably is. For what else would you expect, if consciousness is itself a form of mystery-laden matter? Maybe if it were made of immaterial God-stuff it might yield more readily to our powers of understanding: but if it belongs to the world of gravity and electric charge and invariant light speed and quantum paradox, it is not a bit surprising if it refuses to yield up its secrets.[17] Those problems are *hard*. If consciousness is indeed a form of matter, then we would *expect* it to belong in the class of natural mysteries, not to be a model of epistemic transparency.

---

[16] As you know, consciousness is already a member in good standing of the Mind Association. But I gather the Matter Club is usually regarded as a more prestigious institution. See chapter 8 for more on this (the Matter Club is run by skeletons, apparently, while the Mind Association has flesh and blood members).

[17] At least in the case of God we can somewhat model his being on ours, because he is a conscious subject of sorts (but with no attention problems). So we have some empathic understanding of his nature. But in the case of the material world we are dealing with an alien form of reality, relative to our mental nature, whose inner being has only been revealed to us slowly and painfully (and partially). God is actually more intelligible to us than matter, once you delve more deeply into the epistemology of the two (substitute "the gods" for "God" in this statement, if you like). And immaterial substance, being really a product of human imagination, is bound to be more intelligible than material substance, whose objective nature we must strive to discover. We generally know our merely intentional objects better than we know real objects, because there is no more to them than what we put in. Fictions are easier to grasp than facts, because they are a matter of stipulation. God is human subjectivity writ large, and without the kind of material body whose nature is so mysterious to us: so we have a better chance of knowing his nature than we do of knowing the nature of ordinary material things. For much the same reason ghosts are easier to comprehend than real people—being fictions, they are ontologically thinner.

Someone might object to my whole procedure in this paper, as follows: "You say that consciousness is a form of matter, but you have said nothing about the nature of matter. Specifically, you have said nothing about *how* matter might take this form, as well as other forms. You have therefore done nothing to enable us to *understand* how matter can take the form of consciousness. In fact, I hear that you yourself don't even think we have any positive descriptive conception of matter. So how can you claim with any confidence that matter is inherently capable of taking the form of consciousness, in addition to its other forms? Isn't the intrinsic nature of matter quite hidden to us?" I have a good deal of sympathy for this objection, and indeed it represents my views perfectly correctly. I have certainly not claimed that the essence of matter is such-and-such and that therefore anyone can see from this that matter is capable of taking the form of consciousness. I have merely brought up some methodological and circumstantial reasons that favor the hypothesis that consciousness is a form of matter. In order to know with certainty that the hypothesis is true, we would need clear and distinct ideas of matter in its objective essence, so that we could see the seeds of consciousness there—which I think we are very far from possessing, to put it mildly. But I also think that nothing we know of matter rules out the hypothesis, and there are indirect reasons for favoring it—so we can certainly rationally entertain the hypothesis. Still, our lack of insight into matter surely conditions our sense of whether consciousness might be one mode of matter, perhaps leading us to resist the idea. This appears to be one of those frustrating cases in which we can apparently know that something is true without at all understanding how it can be.[18]

---

[18] As I have had occasion to point out in the past, such situations arise when deep-seated cognitive closure obtains. We are then in the position that we know that something is so without being able to grasp what makes it possible or what explains it. A moderately intelligent ape might know quite well that the sun rises every morning or that bananas are nourishing, but it does not follow that he or she is capable of learning or grasping what makes these things the case—orbits and digestion, to put it simply. Knowledge of facts does not always bring with it knowledge of the explanation of facts, even in principle. We humans might be thus limited when it comes to explaining how matter can assume a conscious form. After all, we too are just moderately intelligent apes. See my *Problems in Philosophy: the Limits of Enquiry* (Basil Blackwell: Oxford, 1993).

11 Matter and Meaning

Physics has long distinguished itself as the exemplary science, admired and respected. Other disciplines have envied physics and tried to model themselves on it. The social sciences, in particular, have felt a strong urge to mimic the methods and techniques of physics. Its manifest success has exerted a huge impact on intellectual culture in general. Physics is the shining emblem of man's intellectual powers, his ascendancy over nature, his godlike status. When the physicist speaks, the world listens. Physics is, above all, *influential*. In this paper I shall explore the influence of physics on philosophical work relating to language and meaning—one area in which the intellectual preeminence of physics might make itself felt. We shall find some interesting parallels. I begin with an historical preamble.

The history of physics shows an interesting arc. If we date its beginnings to Greek atomism, we already see some key features of later developments. Matter is held to consist of relatively homogeneous indivisible elements—atoms—that exist in the "void" (empty space) and that combine together to produce observable bodies. The atoms themselves are unobservable, but they produce observable things by moving around in space to form new configurations. Matter is thus conceived as corpuscular in structure, with its discrete constituents represented as combinatorially inclined: somehow the atoms stick together to form larger wholes. At this stage, the idea of force is not yet fully present, and the particles are pictured as inert and inactive in themselves, having motion imparted to them by some outside agency.

The ancient atomism was taken up by the mechanists of the seventeenth century. Descartes thought of matter as consisting of discrete units characterized by extension and motion—an even more exiguous basis than the Greek version—and interacting by means of contact causation. The world is an elaborate machine, made of mobile parts, the essence of which is extension. Force, again, is deemed secondary to the idea of material constituents that combine together. Newton's physics is essentially the same, except that he introduced gravity as a force that acts at a distance—which was anathema to Descartes and his mechanistic followers. Newton added his quantitative laws of motion, unifying terrestrial and celestial motion, but the underlying mechanism of Descartes is preserved—masses in motion, matter as corpuscular, force as extrinsic to matter. The notion that matter is punctiform at the microscopic level is preserved by the assumption of point particles that aggregate into complexes, being simple and indivisible in themselves. Things resolve into their atomic parts if we dig down deeply enough, and there we find a simple and unchanging order. It is the hidden army of discrete individual atoms that makes the clockwork of the universe turn.

This model began to be challenged when physicists turned to light, magnetism, and electricity. Here the notion of a *wave* proved central, a notion quite distinct from that of a material corpuscle. A wave is spread out in space, is not solid, does not exclude other entities, is inherently active, and is not identifiable with its physical substrate: it is a concentration of energy, characterized by amplitude and frequency. A wave is essentially force-like or energetic. Maxwell's theory of electromagnetism introduces the idea of a field of force, wherein such waves have their being—a region of space in which the force operates. There are no sharp discrete boundaries to a wave in a field, just a gradual falling off in efficacy (subject to an inverse square law). Waves are not at all like classical particles—they are patterns not parts. The theory of electromagnetic waves was initially pursued separately from the theory of the constituents of matter, so that light and matter were ontologically opposed. But the idea that all of nature was structurally corpuscular had been questioned: physics now had to recognize two very different types of entity. No longer was energy or force the poor cousin of matter, with its hard and respectable rigidity—it was now its equal partner. The essence of the physical world had now to include *activity*.

It was left to quantum theory to raise the profile of energy and force still further (with a little help from relativity theory). There were, before the advent of quantum theory, the "dynamists," like Boscovich, who insisted that the notion of a field of force was basic, with so-called particles being nothing other than tiny repulsive fields.[1] The dynamists rejected the idea of a material punctate particle surrounded by a force

---

[1] Roger Joseph Boscovich (1711–1787) was a Croation Jesuit polymath famed for his accomplishments in physics and astronomy. He made impenetrability crucial to atomic theory, with particles as point-centered force fields, not discrete bounded substances. He is said to have influenced Faraday in the invention of the electromagnetic field. He is generally regarded as a man before his time, theoretically speaking.

field; instead they dissolved the particle into the field, working only with the idea of intensity of force and the notions of attraction and repulsion. The conception of material boundaries to elementary constituents was questioned, and with it the whole atomistic ontology. Instead of nugget-like atoms in the void, nature for the dynamists consists of fields of force spread out in space. Thus nature is conceived as essentially active—dynamic all the way down. Force characterizes the most basic physical reality: it is not an optional afterthought. This general metaphysical picture was reinforced by quantum theory, in which the microscopic entities are subject to "particle-wave duality." Electrons, say, have wavelike properties *and* particle-like properties. Moreover, the concept of energy was applied intrinsically to the hitherto corpuscular entities of microphysics.[2] Einstein had already connected mass and energy, erasing the sharp line that had separated them, but now quantum theory argued that elementary "particles" were really packets of wavelike energy. Not only was energy autonomous, not requiring a material vehicle for its propagation; it turned out to be fundamental. The punctiform picture of discrete particles, inactive in themselves, but moved around by external forces, was replaced by a picture of intrinsically active nodes of energetic force interacting and pervading space. Matter *became* energy, in effect. We might say that the static geometrical conception of matter enshrined in atomism—discrete bounded parts standing in spatial relations—was replaced by a dynamic conception in which action in time became central.[3] The essence of matter is to *do* something in time, not to have a geometrical structure in space. Such is the conceptual revolution ushered in by quantum theory and associated developments. (I don't say that I myself subscribe to all of these alleged metaphysical consequences of

---

[2] Here is a characteristically trenchant passage from Heisenberg: "We may remark at this point that modern physics is in some ways extremely near to the doctrines of Heraclitus. If we replace the word 'fire' by the word 'energy' we can almost repeat his statements word for word from our modern point of view. Energy is in fact the substance from which all elementary particles, all atoms and therefore all things are made, and energy is that which moves. Energy is a substance, since its total amount does not change, and the elementary particles can actually be made from this substance as is seen in many experiments on the creation of elementary particles. Energy can be changed into motion, into heat, into light and into tension. Energy may be called the fundamental cause for all change in the world." *Physics and Philosophy* (HarperPerennial: New York, 1958), 37.

[3] Heisenberg says: "The Greek philosophers thought of static forms and found them in the regular solids. Modern science, however, has from its beginning in the sixteenth and seventeenth centuries started from the dynamical problem. The constant element in physics since Newton is not a configuration or a geometrical form, but a dynamic law" (37). Earlier he states the new ontology of physics thus: "The probability wave of Bohr, Kramers, Slater, meant more than that [subjective epistemic probability]; it meant a tendency for something. It was a quantitative version of the old concept of 'potentia' in Aristotelian philosophy. It introduced something standing in the middle between the idea of an event and the actual event, a strange kind of physical reality just in the middle between possibility and reality" (15). Energy, then, is the dynamic universal substance that consists in potentiality; or better, the nonsubstance that exists as pure active tendency. The static geometric solids of the old particle theory are thus replaced by dynamic powers that consist in *de re* probabilities. Matter does not consist of a pattern of rigid beads laid out neatly in space, but of jolts of unruly energy spread out in time—shoves not shapes.

quantum theory; I am simply describing what the consequences were commonly taken to be.)

The overall arc, in sum, goes from inherently passive structures of simple discrete units in spatial combination to active principles operating over time. The idea of active energy replaces the idea of inert matter. Energy becomes the ontological baseline, the basic stuff. If we think of conceptions of matter as modeled on human experience, then the transition is from the passive geometrical character of visual experience to the effortful character of acts of will: from size and shape to pushing and pulling, in effect. Structure gives way to impetus.

My aim in re-telling this familiar story is not to add anything to it or to contribute to the philosophy of physics. It is to explore an analogy between the conceptual revolution that occurred in physics, as that revolution was understood by the active participants, and the change in approaches to the nature of meaning over the course of the twentieth century. I will be speaking of this analogy at a high level of generality, adopting a lofty standpoint, my aim being to draw out some significant (if hazy) conceptual overlaps in the two narratives. My socio-historical hypothesis is that theorists of meaning were subliminally (sometimes consciously) sensitive to what was happening in physics—the paradigm of the sciences—and this shaped their thinking about meaning. I shall focus on Wittgenstein as the most emblematic, but other thinkers also come into the story.

Norman Malcolm comments that Wittgenstein's theory in the *Tractatus* conceived of sentences (or propositions) as like *mechanisms*, and I think this is a perceptive interpretation (Wittgenstein was, after all, trained as an engineer).[4] A sentence is a combination of signs whose meanings fit together into a complex whole, rather like a machine and its parts. The sentence models reality, as a toy car can model a real car. The meaning of a sentence can be broken down, in principle, into its interlocking components, ultimately reaching simple names that stand for simple objects. These will be revealed by analysis, being normally hidden to the naked eye and ear: there is thus a "microscopic" level of semantic reality. Since the meaning of a name is the object it designates, the constituents of meaning are also constituents of the world; Wittgenstein even speaks of these objects as constituting the *substance* of the world.[5] The simple objects combine to produce more complex objects (as well as "facts"), and hence meanings are built up from simple (indivisible) constituents. Since simple indivisible objects are by definition atoms, the theory is atomistic: each atom is a

---

[4] This is from Malcolm's entry on Wittgenstein in the *Encyclopedia of Philosophy* (Macmillan: New York, 1967), vol. 8, ed. Paul Edwards.

[5] At 2.021 Wittgenstein writes: "Objects make up the substance of the world. That is why they cannot be composite": *Tractatus Logico-Philosophicus* (Routledge & Kegan Paul: London, 1961). Note also the reference to Hertz at 4.04: "In a proposition there must be exactly as many distinguishable parts as in the situation that it represents. The two must possess the same logical (mathematical) multiplicity. (Compare Hertz's *Mechanics* on dynamical models.)" Also at 6.361: "One might say, using Hertz's terminology, that only connexions that are *subject to law* are *thinkable*." Wittgenstein clearly knew his Hertz, one of the most distinguished physicists of his time, and used him.

discrete, ontologically independent entity, but the atoms can aggregate to form the meanings we recognize. Once the combining operation has been performed, we have a sentence with meaning—nothing further needs to be added for sentences to have meaning.

A logical analysis, such as Russell's theory of descriptions, can reveal the hidden fine structure of propositions, so that the basic semantic components stand forth. But even if we have no such theory to hand, we know that a basic analysis must exist, since meanings must rest upon simple semantic atoms. Indeed, in the version of Wittgenstein's theory advanced by Russell, the analogy with atomism in physics is explicit: "The Philosophy of Logical Atomism." Here Russell speaks of "atomic propositions" and "molecular propositions" and "logical particles," expressly supposing a combinatorial picture founded on simples. What these simples are precisely is not made clear by Wittgenstein—are they physical atoms or points in a sensory field?—but it is supposed that atoms there had to be, for meaning to be possible at all. Meaning is at root corpuscular, on this conception, and the mechanism of sentence meaning is aggregation. All atomic propositions are deemed logically independent of each other—rather as physical atoms are deemed separable from anything they might be part of, as well as from other atoms. Semantic reality is an arrangement of particles of meaning in varying configurations.[6]

Accordingly, Wittgenstein sought to minimize the range of irreducibly various entities in his system, regarding meaning as fundamentally homogeneous—just consisting of simple objects in combination. As Newton confined himself to masses in motion, so Wittgenstein recognized only names and assertions: every sub-sentential expression is really a name, denoting an object, and the only sentences countenanced are assertions (imperatives and so on are just disguised assertions). Just as the physical atomists kept their primitives to a minimum, so did Wittgenstein, letting the work of generating variety be performed by iteration and disguised appearance. Really, the world is simpler than it appears, once you get down to the foundations, according to the logical geography of the *Tractatus*. Meaning has an orderly crystalline structure, consisting of logically aggregated atoms. It is precise, articulated, and mechanistic.

The *Tractatus* is strikingly geometrical in a number of respects. First, the very structure of propositions is geometrical: they are pictures of facts—diagrams, in effect—and therefore stand in a relation of isomorphism to facts. Lines of projection can be drawn from language to the reality. True, they are logical, not spatial, pictures: but the prevailing metaphor is certainly geometrical—sentences are *like* geometrical figures, only not strictly geometrical. The very idea of pictorial form, of which logical form is a special case, is geometrical in origin. And the notion of "logical space" is also geometrical: a proposition carves out a region of logical space—the region in which it is

---

[6] These particles fall into groups analogous to physical particles: names, predicates, and connectives; and electrons, protons, and neutrons. A sentence is made of a combination of these particles, as an atom is made of its particles; and in both cases a mechanism of cohesion is necessary.

true—and that region constitutes the content of the proposition. Propositions are *quasi*-spatial, casting a shadow over logical space. So, too, the idea of building meanings up from simpler components has a geometrical inspiration (as Greek atomism surely did), because in geometry we build complex figures up from simpler elements, right down to lines and points (perhaps the inspiration for the notion of a physical atom). The *Tractatus* even contains diagrams that purport to reveal the inner nature of propositions. Truth tables themselves are diagrams with geometrical form. Geometry runs through the entire work, shaping its basic theoretical outlook.[7]

And there is an important historical connection to geometry. It is not an accident that the *Tractatus* is laid out axiomatically, like Euclid's *Elements*, with many a deduction and corollary. Newton's *Principia* was explicitly modeled on Euclid's book, with diagrams, deductions, and theorems, aiming to do for physics what Euclid had done for geometry—it has a markedly geometrical slant. Russell and Whitehead's *Principia Mathematica* was modeled on Newton's book, as is evident from the title, aiming to do for mathematics what Newton had done for physics. Wittgenstein, of course, studied with Russell and was very familiar with his teacher's Big Book. The *Tractatus* is clearly modeled on that book, in that it proceeds axiomatically, trying to do for the theory of meaning (and hence philosophy) what Russell and Whitehead had done for mathematics. So, transitively, Wittgenstein's *Tractatus* was shaped by Euclid's *Elements*, via Newton's *Principia* and Russell and Whitehead's later *Principia*. The intellectual ideal represented by Euclid was aspired to by each of his successors (with mixed success); and the kind of atomism implicit in Euclid's geometry also had its influence—geometrical simples combining to produce ever more complex geometrical wholes. It is the picture of the world as an assemblage of primitive components systematically linked.

The general point I am driving at, then, is that the *Tractatus* contains a "corpuscular" theory of meaning, in somewhat the sense that the term is used in physics. Or, to elaborate on Malcolm's comment, the theory is a type of mechanism—a theory of primitive parts that lock together to produce working wholes. Both matter and meaning are constituted by collections of cohering atoms, of one kind or the other. What should be particularly noted are the abstract structural aspects of the theory expounded in the *Tractatus*—the kind of architecture it ascribes to language and meaning—as well as the absence from it of anything relating to what we *do* with language: its active employment in human communication. Meaning is essentially an abstract structure of elements, divorced from human action—Euclidian, Platonic, Newtonian. A very similar conception permeates Frege's theory of sense and reference. Again, we have the combinatorial picture, the typology of primitive constituents (names and "concept words"), the notion of senses as units of meaning that

---

[7] An analogy is explicitly drawn at 3.411: "In geometry and logic alike a place is a possibility: something can exist in it." This is surrounded by several sections discussing "logical space." There is quite a bit of discussion about mechanics in 6.341–6.3432. Wittgenstein was clearly steeped in physics and geometry.

are discrete and clearly bounded. Any idea of actual human speech ("pragmatics") is strictly extrinsic to this structure of abstract entities. There is also the idea in Frege of logical analysis and misleading surface appearances—with an underlying level of orderly, precise units of sense. Frege even regarded senses as timeless and unchanging, much as the atomists think of their basic units of material existence.[8] In both early Wittgenstein and in Frege, then, meaning is conceived according to the dictates of an atomistic perspective—a fundamentally Newtonian (and hence Euclidian) picture of the kind of thing that meaning is. Nor is that a bit surprising, in view of the enormous intellectual prestige of Newton's work at the time they were writing. How could one *not* have one's intellectual outlook shaped by such a monumental achievement? We might aptly refer to this era as the age of "Newtonian semantics," speaking very broadly, analogously to Newtonian mechanics.[9]

So powerful is this conception of meaning that it is hard to envisage any alternative—just as it was hard for people to see that the old atomistic picture of matter had serious flaws. Replacing matter by energy (to put it very simply) took a severe intellectual effort of rethinking.[10] But what might replace the corpuscular theory of meaning? What analogue might there be for force and energy? Where is the new semantic dynamism to replace the old semantic mechanism? For that we need look

---

[8] Frege was also exercised with the problem of propositional coherence—how senses combine into a unitary thought. This was a longstanding concern of physics too: if bodies are composed of discrete particles, what keeps them from falling apart? What is the basis of physical cohesion? It was a long time until electricity was determined to be the requisite glue. Frege invoked the idea of being "unsaturated," itself drawn from chemistry. Chemistry, being continuous with physics, was an allied source of influence: molecules as complex geometric structures, picturable things, combining and constructing, with a clear internal analysis.

[9] I observe that Frege, Russell, and Wittgenstein approached language and meaning as mathematicians, just as Newton approached physics and astronomy as a mathematician. The power of mathematics to bring order and clarity must have been firmly before their minds. Arithmetic (as well as geometry) consists of a domain of elements that generate new elements by the iteration of primitive operations, e.g., addition. The atomist picture is present in mathematics too.

[10] The following passage from Heisenberg gives a sense of the degree of rethinking necessary in the new physics: "Let us discuss the question: What *is* an elementary particle? We say, for instance, simply 'a neutron' but we can give no well-defined picture of what we mean by the word. We can use several pictures and describe it once as a particle, once as a wave or as a wave packet. But we know that none of these descriptions is accurate. Certainly the neutron has no color, no smell, no taste. In this respect it resembles the atom of Greek philosophy. But even the other qualities are taken from the elementary particle, at least to some extent; the concepts of geometry and kinematics, like shape or motion in space, cannot be applied to it consistently. If one wants to give an accurate description of the elementary particle—and here the emphasis is on the word 'accurate'—the only thing which can be written down as description is a probability function. But then one sees that not even the quality of being (if that may be called a 'quality') belongs to what is described. It is a possibility for being or a tendency for being. Therefore, the elementary particle of modern physics is still far more abstract than the atom of the Greeks, and it is by this very property more consistent as a clue for explaining the behavior of matter" (work cited in note 2, 44). The world has gone from the solid, clear, and articulated to the amorphous, unclear, and confused—yet the new scheme is descriptively more successful, given the data. The preconceived idea of articulate structure has given way to the empirically informed idea of chaotic and ambiguous activity.

no further than Wittgenstein's *Philosophical Investigations*. I noted that the *Tractatus* is silent on the question of what we do with our sentences, construing them as imbued with meaning independently of any actual employment (so long as the machine has been correctly assembled). In the old physics matter was conceived as self-sufficient prior to anything it might do, even move: its essence, declared Descartes, was mere extension, with any motive forces superadded, not part of the inner reality. Similarly, a sentence has meaning so long as the laws of semantic combination have been obeyed, quite independently of anything we might do with it.

But in the *Investigations* meaning is *use*—what we *do* with our sentences, the aims we have in uttering them, the games we play with them, the forms of life into which they are woven. Sentences, in other words, are conceived *dynamically*—in terms of their role in temporally extended human action. They are not static structures whose meaning is fixed at any given moment of time; they are employed over time and their meaning depends on their history of use (customs, practices).[11] Consider those builders Wittgenstein describes: utterances in their primitive language-game produce actions and are actions themselves; they are not the mere parade of quasi-geometrical replicas of chunks of reality.[12] Wittgenstein emphasizes linguistic communication, which is a dynamic interaction (a little like action at a distance, as some have remarked). He also compares words to tools, which are simply instruments through which our actions impinge on the world. An utterance is like a center of energy (a propagating wave) in that it has effects beyond itself—it does "work." It sends out a kind of force, instead of just sitting there inertly, like lifeless matter (at least as matter was conceived in the mechanistic tradition). Meaning is now an activity, not a substance—more process-like than thing-like. Its framing element is time, not space (logical space). Meaning is like playing a game of chess, not like the board and pieces themselves (as it were, before the game gets started).

It might be objected that there is no real conflict between this emphasis on use and the earlier atomistic conception of meaning—we can have both. We simply need to make a distinction between sense and force, as Frege decisively did. Sense is constituted antecedently to use by atomistic principles, but then we endow this sense with a certain (illocutionary) force that gives it the kind of impact (note the physical term) that it has

---

[11] See especially sections 179–242 of *Philosophical Investigations* (Basil Blackwell: Oxford, 1958), in which the active and temporal nature of use is emphasized, as opposed to the static structures of the *Tractatus*. In section 198, for example, Wittgenstein says: "a person goes by a sign-post only in so far as there exists a regular use of sign-posts, a custom." It is natural to feel, reading these sections, how we feel when reading about the new physics: the old reassuring atomistic world-view has gone from meaning, to be replaced by something shifting and evanescent, elusive and infuriating—just a web of human tendencies ("forms of life") with no solid foundation. Wittgenstein and Heisenberg are both trying to persuade us to stop thinking in ways we find immensely attractive and natural—we must be "cured" of our habitual mechanistic atomism. (Heisenberg himself must have had his own "*Tractatus* period" of comfortable mechanism, before the conceptual revolution hit.) The tone of both writers is accordingly both hectoring and apologetic (though subtly so).

[12] See sections 2 and following of *Investigations*. It is interesting that this is a language game in which utterances bring about the *motion* of stones—as if they had kinetic energy.

in communication—either as an assertion or a command or a question. We thus divide an utterance into a "sentence radical" and the force with which that radical is deployed in communication; and the radical is the atomistic structure envisaged by traditional theorists. It is quite clear to me that this division reflects a similar attempt to segregate two different (and opposed) physical realities: the kernel of substantial matter, conceived as corpuscular extension, and inherently inert, on the one hand; and the surrounding forces that somehow emanate from it, notably gravitational force, on the other. In other words, we allow for a dichotomy of matter and energy, giving energy its due, but we don't permit energy to swallow up traditional substantial matter. Similarly, we may hope to register the active ("pragmatic") aspects of meaning while retaining the distinct category of atomistic sense, by invoking a sense-force distinction. And that is certainly an intelligible maneuver, as was the comparable move in physics. We acknowledge the importance of force, but we limit its ontological reach.

However, it was just such a halfway measure that failed in the case of physics, at least according to the received story: matter turned out to *be* energy, or to be so closely linked to it that any hope of ontological insulation was hopeless. Physics abandoned the matter-energy distinction, as a piece of metaphysics, and construed physical reality as consisting solely of energy fields. And I think it is very significant that Wittgenstein abandoned the analogous distinction in his later account of meaning: there is no notion of the invariant sentence radical that is endowed with various forces in the production of speech acts in the *Investigations*; there is just a series of *sui generis* types of utterance.[13] Nor is the later Wittgenstein much concerned with questions of compositionality, and hence with the notion of semantic simples. He simply considers whole utterances and their dynamic interactive properties. His position is analogous to those thoroughgoing dynamists I mentioned earlier: it is force all the way down, so to speak. There is *just* the use, with no background of atomistic semantic structure. Thus what some have seen, not unreasonably, as Wittgenstein's pragmatic turn in the *Investigations*, I locate against a broader background—bringing in large upheavals in physics and not just philosophy. Indeed, philosophical pragmatism itself might be seen as a reflection of the shift within physics from Being to Doing. Or perhaps we should say, less baldly, that meaning has *some* particle-like properties, presenting a corpuscular appearance at least, but that its underlying essence is force-like: any atomistic properties it displays are derivative from use, not fundamental realities from which use derives (with the enrichment of illocutionary force). As a question of deep metaphysics, matter is really energy, it is thought, though it can display particle-like properties. Similarly, as a question of deep metaphysics, meaning is really use, though it too can appear particulate, simply because words can combine to form phases and sentences.

---

[13] See section 23 of *Investigations*. He here asserts the multiplicity of types of utterance and does not even pause to consider whether there might be a semantic constant running through them (the sentence radical or Fregean thought or "propositional content").

Our *picture* of matter changes under the new dispensation, though we can say many of the old things; similarly, our picture of meaning changes, though we can still assert such truisms as that sentences are made up out of words. What we don't have any more is a heavy foundational ontology of discrete units of semantic reality ("objects," "senses," "ideas") becoming attached to each other as words are combined into sentences. There *is* no machine (mechanism) for the ghost of energy or force to be in, and the ghost is actually not as spooky as he at first appeared to be (though it may take a lot of philosophical work for this to become apparent). What we must primarily be made to see is that we don't *need* the semantic machine, though we may think we do—just as it took physicists a long time to accept that they didn't need Cartesian mechanism (or even Newtonian).[14] (I am again not endorsing this view of physics or semantics, just describing it: this is how practitioners and observers saw things.)

I think we can see this dialectic working itself out in regard to changing attitudes towards vagueness. Frege couldn't tolerate it, finding it incoherent, and the early Wittgenstein had no truck with it either: concepts must have sharp boundaries, and meanings must be precisely delineated. This very terminology suggests a parallel with analogous questions about the structure of matter. For an atomist, every particle is a discrete unit with determinate boundaries: there cannot be any gradual fading out of a particle—you are either inside it or outside it, with no intermediate position possible. But if a concept is like an atom, it too will have sharp boundaries: its extension will be defined in an all-or-nothing fashion (note the analogy between "extension" in the logical sense and "extension" in the material sense). By contrast, a wave or a force field is not sharply bounded: it fades out over distance, extending theoretically to infinity. Energy is not spatially circumscribed as old-style matter was supposed to be, with a particular size and shape. The question of where an electron starts and ends in space becomes moot—a vague matter. In the *Investigations* Wittgenstein takes a markedly relaxed attitude towards vagueness, finding it quite meritorious in its way, as well as indispensable.[15] No longer do concepts need sharp boundaries and clear definitions, and their vagueness is no count against their integrity or reality. So the move to the blurry and fuzzy in physics has its counterpart in

---

[14] Here the notion of *potential* becomes crucial: to be a physical particle or word meaning is to have a potential for acting in a certain way—not to have a specific structure or constitutive substance. I think Heisenberg hit the nail on the head when he invoked Aristotelian "potentia" (see note 3)—reality as possibility. I also think Wittgenstein is grappling with the same basic issue of potentiality in his famously difficult passages about machines: sections 193–94 of *Investigations*. Meaning, for him, is not use per se, but *potential* for use—and this is a metaphysically problematic idea (in what sense is the use "contained" in the meaning or understanding?). Potential is the ghost to mechanism's machine—the slippery and intangible something that hovers between the actual and the merely possible. How can mere potential be at the foundation of reality, either physical or semantic? Don't we need something more fixed and determinate, more grounded, more graspable? Can we really replace the machine with the ghost?

[15] See sections 88 and following of *Investigations*.

the philosophy of language. Vagueness in semantics is acceptable because vagueness in physics is: so long as a vague word does something useful, its failure to measure up to atomist prejudices does not count against it—and the same for electrons and the like. If God is not a precise Euclidian geometer in the physical world, neither do we need to be in the linguistic world.[16]

Wittgenstein was not the only one to be moving towards a more dynamic view of language; Austin was moving in a similar direction, emphasizing the speech act and what we do with words. Austin indeed wrote a book entitled *How to Do Things With Words*, and his notion of the performative utterance is very much the idea of a use of language in which some action is performed—some "work" is done (like promising or resigning or marrying). He was against the static image of meaning as a mere mirror of reality, and wanted to bring out the interactive dynamic aspect of speech. Ordinary language philosophy, so called, was in fact moving in the same general direction as theoretical physics, ironically enough. Grice, too, with his account of meaning as audience-directed intention, was highlighting the way meaning affects hearers: to mean something is to try to have an *effect* on a hearer. Intentions themselves are like internal forces promoting action, and they lie behind the very nature of meaning.[17] Quine presents an especially instructive case, because of his thesis of semantic indeterminacy. There can be little doubt that that thesis was inspired by the precedent of quantum indeterminacy: there is no fact of the matter about the location of an electron, it was claimed, and likewise there is no fact of the matter about what words mean (about their semantic location).

Both indeterminacy theses fly in the face of common sense and traditional theory, but the former thesis at least had the imprimatur of established science (supposedly). If physical reality can turn out not to have the fixity we supposed, then surely semantic reality can too. If this is a blow for traditional atomism, in both areas, then so be it. Quine also, in his way, wanted to locate meaning in use, removing it from the realm of a structured substratum—the Fregean world of well-defined discrete units of sense (bits of semantic substance, so to speak).[18] These trends all echo

---

[16] Again, the influence of geometry is apparent in the old attitude towards vagueness, in both physics and semantics: there can be no blurring of boundaries in Euclidian figures, so how could matter and meaning admit of such blurring? Tolerance of blurred boundaries required liberation from Euclidian preconceptions—we had to stop thinking of the world on the model of Euclidian forms, with their sharp lines and clear discontinuities. The world consists rather of interpenetrating fields that grade off to infinity, and meanings that have no sharp boundaries or inner unity. It's all just a lot *messier* than we thought. (Euclid kept a clean house, you can say that for him.)

[17] Electrons and protons differ from each other by the *effects* they have, in the reconceptualized physics—as assertions and commands differ in the effects *they* have. Don't try to gaze, as if through a microscope, at the inner architecture of these things in order to grasp their essence; look, rather, at their interactive powers—at their efficacy, their upshot. They are, as it were, the distillation of their effects.

[18] Quinean behaviorism is an indirect outgrowth of operationalism in physics—the injunction to conceive the thing in terms of its measurable effects. Quine, of course, is a philosopher well acquainted with physics, though perhaps a physics distorted by positivism. So far as I know, he never

parallel trends in physics, despite the distance in subject matter; and that is not at all surprising, in view of the intellectual saliency of physics in the twentieth century—not to speak of the desire to model philosophy/linguistics on a discipline so spectacularly successful. The physical world was dissolving into a fluid, dynamic, energetic system—away from hard packets of bounded matter; and at the same time, the realm of meaning was coming to be seen as active, interactive, and difficult to pin down—away from the old picture of tight little semantic units strung logically together into self-subsistent wholes. Empirical physics had liberated the imagination away from Cartesian-Newtonian mechanism; and the new imaginative universe was available for cooption by disciplines seemingly remote from physics—but with their eye firmly on it. Euclid and Democritus had had their day; now it was the turn of Heraclitus.

If I were to attempt to identify in the broadest terms the key intellectual revolution in these twinned histories, I would say it was the following: the rejection of the dichotomy between Being and Doing (rightly or wrongly: I am just describing). We tend to think that there is a firm distinction between what something is, or what it is composed of, on the one hand, and what it does or how it operates, on the other. Matter was the first to experience the repudiation of that dichotomy, and meaning followed soon after. Particulate structure is not the crucial issue here; it is the question of whether the basic constituents of reality are passive or active in their inner nature. Is matter intrinsically inert, needing an outside force to become active, or is it active in its very essence? Is meaning a passive image or reflection of reality, which needs no human employment in order to be what it is, or is it inherently action-involving? Is the *being* of matter/meaning bound up with the *doing* of matter/meaning? On the old view, existence precedes agency; on the new view, agency constitutes existence. Wittgenstein was fond of Goethe's dictum "In the beginning was the deed," finding in it a pithy statement of his own later approach to meaning. But that slogan could just as well sum up the approach to the physical world that replaced mechanism: physical reality consists essentially of dynamic deeds, not inert substances—for its essence is energy. The *act* is the fundamental concept in both areas, not the object. A panpsychist who favors the older conception will tend to identify matter with sensation, sensation being passive; but a panpsychist proponent of the later conception will tend to identify matter with will, since will is active. Perhaps, indeed, our very notions of active and passive stem from this contrast in our mental life: this is the primitive basis of our grasp of those highly general categories. Certainly, many

talked about the structuralist position and its implications; and I doubt he ever referred to Russell's *The Analysis of Matter*. He gives the distinct impression of finding in physics a refuge from the messy and unknowable—the very opposite of what I am advocating. What would he think of Heisenberg's despairing description of the atom? It sounds a lot like Quine's description of meaning! Perhaps he was essentially a Cartesian materialist at heart. Certainly, physics presented itself to him as an epistemological ideal, untarnished by ignorance and doubt.

defenders of physical dynamism have openly celebrated the analogy with the human will, finding in that doctrine a vindication of the divine will in nature.

In any case, the replacement of old-style matter by new-style energy is essentially a replacement of the passive by the active. In some formulations this is tantamount to the contention that matter (if we are to persist with the term) isn't a *substance*—a solid enduring chunk of material, roughly. The world of matter is now conceived as a series of *events*, not persisting things.[19] By analogy, we could say that twentieth-century philosophy of language decided that meaning isn't a substance either (recall that the early Wittgenstein explicitly assimilated meaning to substance, holding that the objects that were the meanings of names constituted the substance of the world). There are events of speech in this *Weltanschauung*, but no substance-like particles of sense. Neither matter nor meaning rests on a foundation of unchanging substantial simples, the permanent reality behind the changing appearances; instead, as the ancient sage proclaimed, all is flux. Matter and meaning are now *defined* by their propensities to action—that is, dynamically. The deed is all. Action replaces configuration.

The topic of semantic change followed naturally in the history of twentieth-century reflection on meaning: for if meaning is a function of what we do, can't we do it differently? Where use changes, meaning must—and there is no solid substratum of semantic substance to keep meaning on the rails. The rejection of semantic atomism thus had far-reaching consequences for the constancy of language over time.[20] There is nothing outside of use to tie use down, to make one use right and another wrong—no hard reality of meanings (such as Frege supposed in his theory of

---

[19] This is Russell's preferred way of stating the metaphysical outlook he takes to be implicit in modern physics: see his *The Analysis of Matter* (Dover Publication: New York, 1954), esp. chapter XXVII. He was passionately opposed to the idea of substance or persisting thing, which he took to be self-evidently absurd. I do not share his passion: I am happy with both substances and events. I don't accept the idea that the world consists merely of events. For one thing, without persistent objects there can be no such thing as motion, since motion is precisely the *same* object being in different places at different times (or in a different relation to other objects). In an ontology purely of events, nothing moves: events of moving need an entity that moves, because moving is the translation of an object to a new place—not the occupation of distinct places by distinct objects (in such a world nothing moves). Oddly, Russell seemed comfortable with persistent propositions: the same proposition continuing over time, perhaps believed or expressed by different people at different moments. He had no tendency to reduce propositions to events (compare Frege on Thoughts). He believed in propositional substances but not physical substances, in effect. Wittgenstein, for his part, did not share Russell's antipathy to persistent physical substances, but he was less sanguine about the notion of temporally continuous meanings, which he found problematic.

[20] There is an important disanalogy between meaning and matter in respect of constancy over time: basic physical quantities are conserved, but there is no such thing as semantic conservation. Energy will transform into different forms while remaining constant, but when meaning transforms nothing basic is left unchanged. There is not a fixed amount of meaning in the universe! Meanings can alter dramatically with use as the arbiter, with no underlying constancy. We could all cease meaning anything by our words if use became sufficiently random, if our customs turned to chaos. But energy cannot be destroyed by anything: the universe is stuck with it. Semantic change is not similarly backed by semantic constancy.

immutable senses). Once we accept that there is no Being over and above Doing—once the deed is made foundational—the whole intellectual landscape changes. Now meaning is essentially a product of human decision, not mind-independent structure, abstract or concrete. Wittgenstein took his adherence to Goethe's dictum as revolutionary, which indeed it was, but in fact the sentiment expressed thereby had long been working its way through physics: it came to apply to meaning only latterly (and probably as a result). Antimechanism about the physical world was the root of antimechanism about the mind and meaning. Or better: once mechanism in physics lost its attractions, the way was open for a like revolution in semantics—semantic mechanism went the way of physical mechanism. But this implied that meanings are essentially changing: they are not robust entities that constrain use, but the upshot of use—a kind of flickering shadow of use. There is nothing to keep us on the rails but ourselves, with our messy and variegated use—with our freedom and creativity. Whole ideologies have sprung from this (putative) perception.

I now want to comment more generally on some other conceptions of the mind that can be plausibly traced back to prior ideas in physics, beginning with functionalism and its ancestor behaviorism. Operationalism as a self-conscious movement only began with Bridgman's coining of the term in the 1920's, but as Bridgman himself noted he was simply following what had already been the norm in physics for some time.[21] The idea was to define physical notions in terms of the operations performed in detecting and measuring the corresponding entities—for example, length in terms of measurement by a suitable measuring rod or electric charge by the amount of motion generated (as measured by distance covered and time elapsed). The upshot is to define *things* in terms of *actions*—as mass is defined in terms of the force it takes to move an object. Mass is as mass does. And what mass does can be measured. Thus mass is defined *as* a measure (in kilograms usually).

The details don't matter for our present purposes: what matters for us is that the behaviorism that started in the beginning of the twentieth century was a manifestation of the same methodological decision. An organism's "drive," say, would be defined in terms of measurable behavioral responses. Thus motivation came to be understood operationally as that which produces a certain behavioral outcome—persistence, basically. A rat had a strong drive to obtain food if it would endure electric shocks in order to obtain the food—where the shocks and the rat's behavioral persistence could be measured (say, by response frequency). The psychologist could then make comparisons of "drive strength" as between different rats, according (say) to how long the rat had been deprived of food. Drive is essentially treated as the physicist treats mass: you find a symptom of it, measure that symptom, and then declare that that is what the original thing is—what is thus measured. Behaviorism, then, proceeds directly from the operationalist methodology

---

[21] See P. W. Bridgman, *The Logic of Modern Physics* (Macmillan: New York, 1960; originally published in 1927).

of physics—quite explicitly. The aspiring psychologist models herself on the admired physicist, with great hopes for the intellectual outcome.[22]

Functionalism, the natural descendant of behaviorism, can also be seen as yet another iteration of operationalism—construing mental states in terms of what they *do*, not what they *are* intrinsically. As mass is defined in terms of (measureable) inertial force, so psychological states are defined in terms of the (measureable) inputs and outputs they "mediate." What this comes down to, in the end, is defining mental states in terms of their tendency to cause motion, because behavior just is a type of motion—which is exactly what the physicists were doing in their operational definitions. Force and energy are operationally defined in terms of motions induced, instead of by any intrinsic features; similarly, the functionalist thinks of mental states in terms of induced motions of bodies. Behavior became central to psychology (and philosophy of mind) because it is a form of motion, and physics defines its key concepts operationally in terms of motion. The behavior of organisms plays the role for the psychologist that the motion of inanimate bodies plays for the physicist—not surprisingly, since behavior is a special case of motion. Behaviorist or functionalist psychology is literally a theory of motion (or would be, if it had got anywhere), and so it defines psychological concepts in terms of motion.[23] I would contend that physics itself is a kind of functionalism about the physical world—black-box input-output science—with the predictable ontological and epistemological gaps: but my present point is that functionalism in psychology is an instance of the same trend that we find in physics (right or wrong). It is the avoidance of the intrinsic features of its subject matter in favor of things it can operationalize—in the end reducing to the measureable motions of bodies. The nature of the ultimate *causes* of such motion is left unspecified, and often unacknowledged.

The doctrine of semantic holism also has a physical inspiration, in the concept of a field of force. If we think of our conceptual scheme on the atomistic model, then we will see each (primitive) concept as independent of the others—as a self-sufficient unit. This is the cognitive particle picture. But if we think instead of a field in space, each point or region depends upon the others: you cannot remove one part of it and

---

[22] Sadly, nothing much came of it, with behaviorism dying a slow and painful death (not that its ghost does not still haunt us). It is an interesting point that psychological quantities, such as degree of belief or intensity of sensation or strength of drive, cannot be measured directly, but only by way of their causes and effects—such as the quantitative character of the physical stimulus or the measurable aspects of the behavioral response. We have no measuring instruments we can just apply to the mental state itself (neural imaging measures by way of brain correlates).

[23] Notice how different this is from defining mental states by way of introspection, in the style of James and Wundt: here motion plays no role at all. And doesn't it seem strange to suppose that the essence of thought, say, is *motion*? What has moving got to do with thinking? It is true that thinking can cause movement, but we often think without moving, and thinking hardly consists in movement. Thinking is indeed a type of action, but mental actions are not a type of bodily movement (they may be types of cerebral movement, as the brain sends its impulses hither and thither). Why would anyone ever have thought it plausible that thinking is the body going from one place to another? Yet this is what the behaviorists vehemently maintained.

leave the rest intact. The field operates as a whole—like a web—and no point in it can act alone. A wave is the undulation of a medium, and undulations take in large tracts of the medium; they are not like discrete corpuscles in the medium (a wave in water is not like an individual water molecule). If we apply this picture to meaning and the mind, then we end up with the idea that concepts are somehow distributed across a medium of some sort (again, I describe, not endorse). They are dispersed not localized, patterns not particles. In fact, the idea of holism is often explained by contrast with something called atomism, and the analogy with fields of force is close to hand. Instead of viewing the mind as an assemblage of separate conceptual atoms, it is viewed as a field in which different waves of force cross and interpenetrate. In this way the concept of a field is used to confer respectability on an idea that might seem flaky: for the concept of a field is a concept in good physical standing. Not every physical reality has detachable sub-units—so why should our conceptual scheme?[24]

It may be objected that the analogies between physics and semantics to which I have drawn attention are just that: they are merely metaphors that might be applied to meaning and the mind, not literally true theories. That is no doubt true, at least to some degree. The notion of a corpuscular semantic substance looks particularly metaphorical. But what does this show? It doesn't show that such metaphors have not guided thinking about meaning and mind. Nor does it show that there is not some more abstract set of ontological categories under which both matter and meaning fall (say, Being and Doing). What is quite true, and often remarked upon, is that metaphors are particularly powerful in theorizing about meaning and mind, whether they are ultimately appropriate or not. Maybe the metaphors of atomism, combination, substance, energy, force, and field are all inadequate, just so much groping in the dark: but that doesn't mean that they have not shaped the dialectic. In the absence of anything more directly applicable, it is not surprising that the paradigm of physics has influenced the way more recalcitrant domains have been understood (or misunderstood). When you can't think of anything very good to say about subject A, it is sometimes comforting to pretend that subject A is just like subject B. The problem is that if the metaphors are forsworn it is not clear what we are left with. We seem bereft of any *non*metaphorical way of talking about mind and meaning, at least in respect of basic theory.[25] We have had theatrical metaphors, homuncular metaphors, mechanical metaphors, hydraulic

---

[24] For the record, I am not an enthusiast of conceptual holism, though I am a great admirer of fields. The concept of the field of force, spread out in space, real yet impalpable, is one of the great innovations of human thought, not least because it defies the dogmas of empiricism. Conceptual holism, by contrast, is standardly motivated by verificationism, which is a version of empiricism. So these are really opposite tendencies.

[25] We can talk fluently about them in the vernacular, since we have a rich "folk" vocabulary for mind and meaning, but when we try to become systematic and theoretical, the dubious metaphors pile up. Our literal statements stay at the surface, but once we venture beyond the surface we lose our moorings. We end up saying things like the mind is a calculating machine or the brain is a community of homunculi and thinking we have a decent theory.

metaphors, telephonic metaphors, computer metaphors, biological metaphors—and so through the long and checkered history of psychological theorizing. In any case, I have argued here that the history of thinking about the nature of mind and meaning bears some striking parallels to the history of thinking about the nature of the physical world. Physics envy has led, understandably enough, to physics mimicry.[26]

---

[26] Since I have restricted myself in this paper to historical description, let me end by briefly indicating my own attitude towards the developments I have described. I think neither phase of thinking about mind and meaning has been satisfactory—not the old mechanism or the new dynamism. Both ways of thinking seem to me too close to paradigms developed elsewhere, notably physics. We need a quite new way of thinking that is neither mechanist-atomist nor dynamist-behaviorist: but what that might be I don't know. And there is, I think, no guarantee that we will be able to come up with anything adequate. What does seem a safe prediction, however, is that whatever physics next produces, in its efforts to overcome its own internal crises, will one way or another find its way into theorizing about mind and meaning—distorted and diluted perhaps, but quite recognizable.

PART TWO

Foreword to *Principia Metaphysica*

The following text is composed in an unorthodox style. This style corresponds to the form in which the ideas occurred to me. It seemed sensible to preserve this form, instead of trying to force the writing into the usual method of exposition. The style is certainly not my normal one—I tend to prefer a pedantic accumulation of explicit argument. I regard this piece of writing as in the nature of an experiment.

It would be pointless to deny the influence of Wittgenstein (as to form not content). Anyone steeped in his work will fall into his patterns once the aphoristic form is accepted. In the present case, the result is a stylistic mixture of the *Tractatus* and the *Investigations*. I have, however, tried to keep the perils of parody to a minimum, though the master's voice creeps in at several points. The problems of pretentiousness presented by the form seem to me insoluble, and I must apologize in advance for any cringes I may induce.

The title is tongue-in-cheek (not to say cheeky). I am *not* claiming any parity with other works similarly entitled. But I couldn't think of a better title, so I let it stand. It might be given the sub-title "Laws and Reality," since this describes the main themes accurately enough.

I recommend reading it all through quickly for the first reading and returning for a slower (and more critical) reading once the general shape has sunk in. The reader is expected to do most of the thinking for his or her self, with my text as a series of signposts. I am not primarily seeking assent from the reader, but rather engagement and self-examination.

As a general point, I think we philosophers have fallen into a style of writing that is rather limited as a means for expressing philosophical thoughts—the conventional journal article or doctoral dissertation (nothing presupposed, heavily footnoted, metaphor free). It is instructive to compare that style with the variety of prose styles that the outstanding philosophers of the twentieth century adopted (as well as earlier): not only Wittgenstein, but also James, Pierce, Russell, Ryle, Austin, Quine, Husserl, Sartre, and others. I doubt that any contemporary philosophy journal would publish the text that follows (possibly justifiably), but I believe it expresses the ideas in a form that suits them perfectly. In any case, I hereby enter a plea for stylistic variety and tolerance.

Principia Metaphysica

1. Objects are composed of stuff ("matter," "energy," "ectoplasm"). This stuff gives the *substance* of the object. Is it a substratum? We can say that, though it need not be unknowable. (Truisms quickly turn into absurdities here.) It is not a priori what *kind* of stuff the world contains, though it is a priori that it contains stuff of *some* sort. The same stuff can take many forms. An object is not reducible to the stuff that composes it. Objects and stuff belong to different "categories."
2. Natural laws are basic. They constitute the form of objects. An object is a nexus of laws. The nature of an object is given by the laws that hold of it. It is an *if-then* world. A particular object is the product of its stuff and its laws. Causality is the expression of laws by means of objects. Laws are invisible, but their manifestations are not. There is no stuff without objects, and no objects without stuff. Laws are conceptually anterior to objects and stuff.
3. Not all states of affairs are constituted by laws and stuff. These may be called "a priori."
4. Laws are not constant conjunctions of objects or events, not even necessary ones; they *constitute* objects (and events). Even if the stuff of the mind were the same as the stuff of corporeal things, the laws would have to be different (same substratum, different nature). Mind and matter cannot share their parts, since parts are objects. Parts are not stuff: even the smallest parts (the basic objects) are composed of stuff (one can always ask "what is it made of?"). The parts of objects are also objects.
5. The ethical might be defined as what we wish the laws were and are disappointed they are not. The meaning of life is ethical, not natural (it cannot be extracted from

the laws). Beauty, however, is determined by the laws of nature (and not just natural beauty). It is consistent for a world to be both evil and beautiful (except when it comes to the soul). Politics arises from the fact that the rules of society are not dictated by the laws of nature. These rules are a matter of convention, not natural law. Conventions need to be justified; natural laws do not. We choose a political order; we don't choose a natural order. Politicians have an interest in conflating the two sorts of order, though it is hard to do so.

6. Ethical truths are not about supernatural objects. There are no such objects, if that means objects that obey no laws. Theological statements purport, incoherently, to be about supernatural objects. God cannot be an object in good standing unless he is constituted by laws. But then God is as defined and limited by his laws as I am by mine, or as a lump of copper is. *I* am constituted by the laws of my "soul." (The laws of my body are not *me*.) Is God wholly constituted by the laws of his soul, in contrast to the body he lacks? The truths of logic and arithmetic, on the other hand, rest on no natural laws; they can survive the extinction of all natural (and supernatural) laws. Not so for the object that is God. Nor can ethics be abolished by the end of laws. But beauty dies when objects and laws do.

7. Even if we do not know whether objects exist, we know that they must obey laws if they do. I may be hallucinating all these objects, but they must obey laws anyway. An unreal world must be law-governed too. Objects cannot fail to be governed by laws, even if they lack existence ("intentional objects"). For laws are essential to our *ideas* of objects.

8. Theological propositions try to be about objects that transcend the natural, but this transcendence would have to hold no matter what natural objects a universe contains. There cannot be a completely supernatural world, because the idea of the supernatural depends upon a contrast (so it is not like the concept of goodness, say). Even incorporeal spirits on earth are not supernatural enough. The substance of the true God, as normally understood, cannot be the substance of the world, no matter what that substance is, including the kind of substance that composes *my* soul. The idea of the substance of God is descriptively empty: it is what is *not* any other kind of substance. The material of objects can indeed be "immaterial"; but that is not enough to make God. God is really conceived as a "limit case" of the natural, a kind of asymptote.

9. What is empirical is the combination of laws and stuff, not each singly. Laws don't depend on objects, but objects do depend on laws. Hume's empiricism got in his way. It takes work not to be an empiricist. Laws and empiricism will never be in harmony. Laws are "entities," though they cannot be seen or touched. Entities that cannot lose bits of their stuff are described as abstract (numbers, say). The abstract world is not supposed to be composed of stuff of any kind, or else it would make sense to think of shedding bits of it. Nor is space a type of stuff for the same reason. The mind, though, can have bits chipped off, just like the brain. A law is not composed of anything: it has no substance.

10. The laws of nature are not unlike Plato's Forms. Yet the laws of nature are the most "concrete" of things. Laws permeate the universe. They are part of empirical reality. Laws do not underlie reality, or transcend it, or correspond to it; they *are* (concrete) reality ("is" of constitution). The reality of objects is inherently general or universal (though not in the trivial sense that the nature of an object is fixed by its properties). The particular cannot be separated from the universal (law). Just like Plato's Forms, laws are both inside the world and outside it—immanent transcendents, so to speak. What is surprising is that *this* object is created by something so far-reaching, and so very different in nature.

11. Reality is also intrinsically temporal, because laws operate over time. The idea of timeless natural objects makes no sense. Space and time are the theater within which laws operate. The laws of space and time are really the laws of their occupants.

12. Laws keep working: they never run out of energy. A law is like an unmoved mover. When a law makes something happen nothing is subtracted from it—there is no depletion of its power. All laws are perpetual motion machines. A law is limitless in its resources. There can be no such thing as a spent law. The basis of causality is therefore, in a sense, infinite. The "counterfactual" character of a law is simply its inexhaustibility. A law is like a bottomless well—as infinite as space and time.

13. There is nothing magical about laws—though they seem awesome. *We* are nothing like laws, though they create us. Human agency is nothing like the agency of a law. Laws are immortal. It is wrong to speak of a law as existing over a period of time. A law must never be confused with its temporal manifestations (yet without time a law would be nothing.) A law can never perform at less than full strength. There is no such thing as a lazy or fatigued law. A law is never less than perfect in its operation.

14. Could God create the laws? Only if he was already governed by laws, and then *they* would be the true creators. God as the "first cause": but the laws that define him are the real causal basis of his actions. Laws are really the first cause, the origin of all causality. "Nomologicism": objects are to laws as mathematics is to logic (roughly). Mathematical objects sprout from logical roots, as natural objects sprout from nomological roots. (Is this a type of reductionism?)

15. Causation lies at the inner core of reality; it is not something superadded. The generality and agency of laws are essential to them, and are imparted to the world by means of causation. The great mistake is to think that laws are ushered in by particular objects—as one thing might be guided by another pre-existing thing. The laws are more like the threads of a garment: no threads, no garment. Even God cannot escape the laws composing his nature (he cannot then exist). For it is part of his nature not to be subject to anything beyond him, even the laws that give him his nature. But laws themselves have no such dependence. Laws are more godlike than God.

16. There can be no causation without laws. A law fixes how objects will behave and interact, *when* they come to exist; thus, their causal powers. There are objects, laws, causation, and properties—but these are not four *things* ("realms").

17. There is nothing psychological about laws—this is quite obvious. And a psychological law has as much efficacy as any law. So-called exceptions to laws are not their breakdown or suspension, but their being overridden by other laws. The effects of laws can be obstructed, but not cancelled. The operation of a law never fails to be felt by the universe. A law cannot be turned off like a light, or even turned down. Varying initial conditions merely set the stage for laws to do their unvarying work. Are some laws more powerful than others—stronger? No. Statistical laws are just as "rigid" as deterministic ones. The laws that govern heaven are just as onerous as those that govern earth.

18. Particulars without laws are empty; laws without particulars are paralyzed. A law without instances is like a disembodied will—impotent. In *this* sense the law cannot precede the particular. Can laws exist without particulars? Whatever the answer is, the important point is that laws constitute particulars. A particular has no efficacy without a law to "back" it (bad metaphor). It is misleading (but not false) to speak of a particular as a cause, as if it could operate on its own, autonomously. All causal statements are really general in nature. Causes are universals (laws not properties—*relations* of properties perhaps). It is not false to say that the law of gravity causes bodies to move as they do. The movements of the bodies are effects, not causes (in the primary sense).

19. If you try to think of natural laws as like legal laws, you do them an injustice. No one lays down a natural law (not even the Universe). Natural laws cannot be repealed, or broken. Legal laws cause nothing, except insofar as they are accepted, whereas the power of natural laws is indifferent to their acceptance. It is odd that we use the same word for both.

20. Nothing concrete can exist without laws; they are the "ground of being." Our senses are not attuned to the intrinsic ontological nature of laws, though we know their existence very well. The foundations of reality are unlikely to be directly revealed to us, though their symptoms are bound to be. Though we do not perceive laws, everything we do perceive is the result of them. The existence of laws is not a "conjecture." Science does not "uncover" laws, if that is taken to suggest a mode of perception; it seeks to state them. All that this means is that laws are not sensible particulars—and who would have supposed otherwise? Nor are laws imposed on the world; they are imposed *by* it. There is nothing optional about them ("conventionalism").

21. The laws of motion strike us as paradigms: is that because motion is so readily detectable by the eye? Would other types of law seem paradigmatic to other types of creature? In motion one sees the laws at work—the fruits of their labor, so to speak. Without the laws of motion objects would not continue in a straight line at a uniform velocity; nor would they swerve or slow down. Am I saying that laws are active and particulars passive? No, because particulars are the vehicle of laws—their "embodiment." It is *in virtue* of the laws of a particular that it can be said to be causally active. You could change all the particulars in the world for distinct ones and preserve the same (causal) laws, but *not* vice versa.

22. In a way this is all a matter of placing the correct emphasis—but then so much of philosophy is. Philosophical advance often consists in a slight change of emphasis. This is why italics carry so much weight in philosophical writing (and why nonphilosophers often can't see their point).

23. I would like to be able to say outright that laws necessarily come *before* everything, even God—but that is not quite right (though sometimes hyperbole serves sobriety). It is as if the laws of the world were the first item on God's agenda, and once they were settled a lot else was too. The laws that govern God are an embarrassment to him, like wearing a low-ranking uniform; he wishes he could cast them off. But without them he is nothing, an empty ego, a frictionless point, a featureless receptacle—a metaphysical vacuum. The laws of God would have to apply to other gods who share his nature: the gods are *subsumed* by their laws. The idea of the supernatural is not scientifically dubious; it is metaphysically incoherent. *Any* object consists of law-governed stuff—so where is there room for the supernatural (in the sense of an object subject to no law—or to "quasi-laws")? Try to conceive of a universe in which *every* object is supernatural. Supernatural compared to what? No object could participate only in miracles, if a miracle is defined as an exception to natural laws. (A law is actually the nearest thing to a miracle that we have.)

24. Laws are produced by nothing, but produce everything. Laws do not impose order on the world, as if the world were a disorderly place till they came along. Can you rely on laws of nature? Not as you rely on the word of a trusted friend. Laws are formative, not merely reliable or useful for prediction. The sun may not rise tomorrow—it may be blown out of the sky by powerful aliens. But that is no abrogation of the laws of nature. To abrogate the laws of nature would be to have no sun to begin with. Obviously, laws do not govern the universe in the way a political party governs a country, and yet this dual use of "govern" invites illusions of independence. It would probably be best to reinvent our entire vocabulary for talking about laws.

25. "Laws + stuff = objects": not such a bad way to put it. "Laws are made manifest in objects and events": yes, but that doesn't mean they acquire reality that way. "Objects instantiate laws": true, but not as objects instantiate predicates (one wants to make a distinction here between internal and external instantiation.) "Objects have laws running through them": better, metaphorically—and how metaphorical is "instantiate" anyway? (Compare: "objects 'respect' laws.") If there were no laws, there would only be raw stuff—and that is impossible. Raw stuff is like the featureless unstructured given—a kind of contradiction. Stuff must come in the form of objects, as thoughts must come in the form of intentionality (rough analogy). Lawless stuff is like James's "blooming, buzzing confusion"—a trick of language. Stuff, objects, and laws come in a seamless package—as consciousness and intentionality do. There is no shaping of a pre-existing reality. (Remember that all analogies have their limitations.) Physical atoms are anything but formless; they are the *parts* of objects—not their stuff. God's three major acts of creation—stuff, objects, and laws—are really just a single act. Conceptual distinctions are not ontological distinctions.

26. Must the laws of nature be simple? Strange question: by what standard? (Of course, our *theories* can be ranked according to simplicity). Could there be five seconds in every day in which the laws of nature break down? Laws cannot disappear, not even for a nanosecond. (An exception to a law is just another law.) Not every world is governed by the same laws, but once installed the laws cannot be varied. A law could cease to manifest itself for a while, but that doesn't mean it has gone out of existence. And what would the resurrection of a law be like? If I destroyed all the metals in the world, there would no longer be any actual instance of the laws governing metals—but that isn't to say that the laws of metals have themselves been destroyed. I find myself wanting to say that laws are eternal like numbers, but they are too woven into the world for that to be quite right.

27. The idea that laws are "contingent on particulars": as if they are merely an addendum to the main story. It is a mistake to think that we "project" whatever we don't see. Ask the objects themselves if we project laws onto them! For the most part, laws are assumed not discovered: consider your knowledge of the results of sitting down. Awareness of laws is animal instinct. Laws should not be "intellectualized." Mathematics is not integral to the notion of law. Not all laws are "scientific." Laws are principles of repeated happenings. Repetition is the essence of law. Boringness is an occupational disease of laws. Theory and law are not interrelated concepts. Laws aren't even presupposed, any more than I presuppose that I have a body (or a mind).

28. Are laws, as I conceive them, "occult"? Not at all—they are part of the most mundane reality. What is the best label for my view of laws—"realism," "objectivism," "productivism," "centralism," "foundationalism," "immanent transcendentalism"? None of these is quite right; and yet labels matter in philosophy (too much).

29. There are unseen realities controlling (i.e., constituting) everything—why does that disturb people so? The notion of a material object is more interesting than many people think—more "queer." (What does this do to "materialism"?) Fear of certain concepts is characteristic of philosophy; we need to conquer the fear. Laws are the absolute bedrock of any world, its formative tectonic plates. If we had "ontological eyes" the first thing we would see would be the laws (I picture them as mile-high dinosaurs with blazing eyes).

30. A law is not a mere regularity among particulars, because any observed regularity can be made irregular. Laws do, however, force regularities to exist. Regularities are signs of laws. We can see regularities, but not laws. A world without repetition is no world at all. Every second billions upon billions of things are repeated. Novelty is not incompatible with law, but it too must be repeatable.

31. Where there is causation there is the possibility of repetition. Causal relations require laws: but not because laws *follow* from causation; rather, laws *are* causation. We could say that laws are the only true causes and be accused of only mild exaggeration. It is not that *a* causes *b and a* and *b* fall under a law; rather, *a*'s causing *b* consists in their falling under a law. The causal ingredient is the law itself. Particulars

can only be said to be causes in a derivative sense—as the transmitters of causal efficacy. Causation is the reflection of law in events. The ability of one event to bring about another depends *entirely* on the law that governs them. Of course, this doesn't mean we always *know* the law in question. If the real causes are laws, it is not surprising that particular causes should always be "backed" by laws; there is nothing adventitious about it. The lawlike character of causes is not something tacked onto them. Laws are the "active ingredients" of the universe, woven into particulars. (This whole subject has been distorted by the "bias in favor of the particular.")

32. Any particular can in principle be destroyed; but laws are indestructible. The first law of laws is their persistence. There has never been even the slightest change in the law of gravity, even as the universe has gone through massive transformations. The unchanging is the source of all change (Plato). Causes don't vary; initial conditions do. (It would be absurd to worship natural laws—though they are rightly feared.) Laws have effects, but not causes. It makes no sense to try to prevent a law of nature from operating. Once a law comes into existence nothing further is needed to keep it operating in perpetuity. It is self-sustaining.

33. "An explanation subsumes the particular case under a law": that is, it says what law *produced* the particular case. When we can't explain something we don't know what law is responsible for it. Not every way we have of describing things identifies the laws that govern them (Davidson). The laws are often not obvious or commonsensical; nor is it always obvious what the *nature* of an object is. But it is hard to know what it would be to be *completely* ignorant of the laws of nature. Language is not automatically keyed to the laws, and thought is not always law-seeking. (These obvious truths are just that.)

34. Events uncork laws, so to speak. We can't see the full causal background of events. An event is the site at which laws invisibly operate. The event of my flipping the switch didn't *au fond* cause the light to go on: for the effect would not have occurred unless it was a law that switch flipping causes lights to go on. An event without a law would be causally impotent. We must not be too slavish about ordinary language here. Did I touch the door or was it my hand? We can say both, but my hand was the more basic thing. Laws are what *power* particulars—give them their causal charge.

35. When God contemplates the laws of nature what sort of sensation does he feel? Perhaps he experiences a sense of wholeness—of connectedness (God is surely a natural monist); and he wonders at his own creative power (forgetting that laws made him). I once saw a massive log holding up a roof: that is what a law is like. (It gave me a sense of insecure security.) A law is not like a repeated theme in music: that is too willful. We seem not to be able to get far beyond metaphors here, yet some are better than others. Is aphorism the best way to express what is so striking and yet so obscure about laws? An aphorism is a bit like a law, after all: condensed, but with large reach. There is majesty in a law—or a line of great poetry. The human significance of laws is hard to put into words, which is why it figures so little in philosophical discussions. Physicists often speak in awe of the laws of the universe (and then

congratulate themselves on having discovered them). I keep returning to the limitless power of laws, their omnipotence. Am I "reifying" them? I am trying to give them their ontological due. If laws were persons we would be duly impressed: should we be less impressed knowing they are not persons? What impresses me most is their effortless productivity. If events are the children of laws, then laws have endlessly extended families. It is interesting that we have a tendency to anthropomorphize laws ("animism"). They are so inflexibly single-minded, so rigid of character, so focused. There is something "authoritarian" about them.

36. Are laws "linguistic"? Clearly not, since there were laws before there was language (nor are they "propositional"). Of course, their formulation is linguistic—trivially. Laws are not entities that sit beside other entities in the world; they *form* those entities (as properties do). The "ontological type" of laws is *sui generis*—which is why they are always being forced in one direction or another. It is not true that laws literally "contain" their manifestations—though a sequence of events contains those events. A law is rather the ground of its manifestations. The same law can have manifestations that look very different (gravity, for instance). Laws unify disparate events. Laws can show up in disguise. Not knowing the laws of motion would be a shame on mankind. All laws are as compelling as the laws of motion. The laws of psychology largely relate what does not move (the mind) to what does (the body): this makes them unusual. Psychological causation involves the transmutation of nonmovement into movement—thoughts into actions. It is exactly *unlike* billiard ball causation.

37. Would it be sensible for God to boast about the power of the laws he has created? No, because that power is intrinsic to them. Nothing *gives* a law power. The power of a law is internal to it. How many examples of laws do we need in order to grasp their essence? Not many—too many would be distracting. The concept of law runs through every law equally (not a "family resemblance"). What is the best way to find out the metaphysical nature of laws? Examine them from every angle. As with most philosophy, there are moments when you seem to have the thing firmly in your sights, and moments when it won't stay still. One would think it would either be clear or not, but it seems to fade in and out of clarity. It is difficult not to think of laws as extremely *big*—the elephants of ontology. Clearly, laws have no size. A "local law" sounds like a contradiction. The law of gravity holds here but not in Denmark? Try to conceive of one law gradually fading out and then being replaced by another. I keep coming back to the image of a law as something like a condensed coil storing enormous potential, like an atom containing so much pent-up energy (laws as universal uranium). But that can't be right (though it *feels* right). I want to *concretize* my idea of a law—which is a weakness. When I am being strict I simply say that everything happens *because* of the laws—but this leaves out so much about the concept. We say what is strictly true in the boring way, but this cries out for expansion—and then the distortions begin. I am trying to strike a chord in the reader (and myself)—even if she doesn't like the chord being struck. Part of my reader wants me to stop, but part wants me to go on. There is certainly something disturbing here, as if the

spooky is gaining a foothold. A law is like a magic ingot of finest gold, pure and unadulterated, rich in potential and possessed of enormous power. It can also be described simply as a principle whereby what happens happens. "Nature and nature's law lay hid in the night; God said let Newton be, and all was light." As if Newton were a prospector who hit it big! Discovering laws makes the universe look a lot cleverer than we thought. Laws seem like brilliant inventions. They never disappoint us ("oh, what a stupid law!"). Mindless intelligence: insentient creativity. One can certainly imagine aliens for whom the existence of laws is the main reason for belief in God. (I am speaking here of the phenomenology of our belief in laws.)

38. The idea that God brought laws to lawless chaos can seem compelling. Laws appear too *advanced* to be taken for granted. "Why is there something rather than nothing?" "Why are there laws rather than chaos?" Except that chaos of *objects* is impossible, and hence impossible *tout court*. Objects are the only things there can be. Chaos is necessarily superficial. One does not take a lawless universe and convert it into a law-governed one by imposing laws on it. The universe cannot be guilty of "disorderly conduct." "The cement of the universe": in many ways a fine image, except that cement goes *between* bricks, without making them what they are. We call them "natural laws," but in some moods they seem quite *un*natural. Do we need to "naturalize" laws? Bad project. Laws are indeed metaphysically remarkable—there is something *fabulous* about them. But remember that things are remarkable by contrast with other things. Life seems remarkable by contrast with inanimate things. But imagine a world of only living things... and then we add some inanimate things. Laws are not remarkable from *their* perspective. Russell's ontological zoo—but the exotic species don't see themselves that way.

39. Would the fact that something had no taste or smell count as a reason to doubt its reality? Whenever something is invisible and intangible we are apt to picture it as a haze—as if a law were like a mist enveloping particulars. It is interesting that we tend to picture the mind this way too. Laws and the mind: objects of suspicion, never quite cleared of all charges; and yet perfectly solid (!) citizens—though perhaps a little "eccentric." Why is solidity taken to be the sole mark of respectability? We must first acknowledge metaphysical oddity, and then insist on its ultimate ordinariness.

40. "The hardness of the logical 'must'": is it really any harder than the nomological "must"? Your house may not be there when you get home, but the laws of nature will be. "Laws shape the world": a potentially misleading metaphor. "Laws are the pith of reality": yes, except that they do not literally lie *inside* objects. In dreams we sometimes escape from laws. Can there be any doubt that we would like to escape from them? Gravity treats us with disdain. Our attitudes towards laws are complex. Laws are implacable, tyrannical: the original master-slave relationship. Accepting the reality of laws is a mark of maturity (powerlessness). In philosophy sometimes all we really need are hints; the rest we can do ourselves.

41. Even the most trivial event is the manifestation of a momentous law. Laws are as endless as space and time. Consider all those different galaxies, and all the atoms in

them—but always the same few laws. (Could this numerical ratio be inverted?) Laws can make the teeming world yet be few in number.

42. The essence of a law is… One feels there exists a perfect completion, but nothing one produces seems to add up to it. Tip of the tongue phenomenon. A law is not bounded, but is it strictly infinite? It keeps repeating ad infinitum. Laws have no natural termination, no given lifespan. They are "self-renewing." The more they do, the more they can do. I find myself looking out of the window as if to catch a law in action. I observe the iron grip of law in every movement. It is possible to *see* the world as a nexus of laws. They infiltrate everything. They are numbingly mechanical. They are both comforting and burdensome. I hesitate to say this, but they remind me of the Freudian unconscious: hidden, yet where all the power resides, with that peculiar combination of the elusive and the ubiquitous—the "id" of the universe. It is so easy to turn them into will o' the wisps, but they are anything but. I am forcing myself to confront the unconscious of the natural world (big "as it were" here). Philosophical poetry is a neglected genre.

43. Science fiction is largely about the human relationship to laws. At some time in the future we have triumphed over them; the master-slave relationship has been reversed. This is a fantasy. The psychoanalysis of laws: breast or penis? Do laws nurture or thrust? The anthropology of laws: boring subject (too universal). We don't *learn* that nature obeys laws. All species have evolved to cope with the same set of laws (so the nature of other species coincides with ours to this degree). Laws are in our bones. Is this why it is so hard to articulate their human meaning? The knowledge runs too deep.

44. What is the minimum number of laws a world can contain? Puzzling question. Whatever is necessary to fix the nature of its objects. Objects can be no simpler than the laws that govern them (and no more complex). Are some laws simpler than others? We would need a measure of simplicity to answer this. One object is simpler than another when fewer laws apply to it (more or less). Laws are no more linguistic or conceptual than objects.

45. Oscar Wilde on laws: "Her hair turned quite gold with grief." The great charm of laws is their utter lawlessness. The working class is subject to the laws of nature; the middle class to the laws of the state; the aristocracy to the laws of God: that is why the aristocracy behaves so atrociously. The chief characteristic of aphorisms is their utter disregard for falsehood. Aphorisms court rejection, cry out for it—unlike arguments.

46. How much point is there in being a "physicalist" once the correct conception of laws is accepted? Are laws physical entities? What point is there in insisting that all of Plato's Forms are physical once you have accepted the reality of Forms? "Every law is a physical law": but are these laws themselves physical?

47. Events can be borderline legal, but not borderline law-governed. Can it be indeterminate whether an event is subsumed by a law? No. Should we then say that natural laws are never vague, but legal laws can be? That sounds right.

48. Our concepts enable us to see things that are hard to put into words. We know quite well when we have not found the right words. Is there something ineffable about laws? Difficult question. One wants to reach the point where there is nothing left to say. It is possible to say it all several times in a single discussion (and usefully). The right thing may need to be said repeatedly (in different contexts, if not always in different words). I want to describe the face of laws as they are constituted when no one is looking. Laws must be detached from their appearance to us, but connected to our real sense of their nature.

49. Natural laws are as pure as logical laws (as "sublime"). Do laws contain their instances in a "queer way" (cf. Wittgenstein on rules)? "Laws constitute the logical scaffolding of the world": no, but the metaphor is natural. Laws relate particulars and properties, not propositions. "Constant conjunction": an excellent description of what a law is *not*. "Laws sustain counterfactuals." Not quite: they are what counterfactuals report. It is not that laws come first and counterfactuals second. Laws extend their reach beyond what actually happens, but this is no kind of bleaching out at the borders of actuality. A law is uniform across its actual and counterfactual instances.

50. Why does metaphysics lend itself to this kind of writing? Metaphysics is "informally axiomatic." There is always more assertion than argument in it. Also, it is inescapably meta-philosophical ("meta-metaphysics").

51. Hume discovered this topic, but then he left too deep a mark on it. We find it difficult to see beyond him. Hume gave an anthropocentric description of laws and causality. He encouraged us to focus on how things look. But laws do not reveal their nature to the senses. "Causality is one damn thing after another": the "after" clearly understates the case, but the "damn" is what really hits the wrong note. The Humean picture of things is, indeed, alluring; one has to fight not to be swayed by it. His view of laws is too flattering to us. There is something bleak (though exhilarating) about the correct view. Not that there is anything heroic about accepting it (resigned perhaps). Laws cannot be cut down to human size. Their *esse* is utterly not *percipi*.

52. One has to be sensitive to one's concepts as to a tender finger. It can be thrilling to gain insight into one's own concepts, strangely enough. Philosophy can give no excuse for its lack of conceptual insight, which is why it is such a chastening subject. I *ought* to know the answer to this! How hard can it *be* to say what "law" means? It is difficult to avoid pretentiousness while discussing the nature of philosophy.

53. An atheist is someone who believes that laws are all that lie behind the sensible world. A theist feels that the laws need external validation. The conviction exists that someone must have chosen the laws. People pray that certain things won't happen, but they don't pray for a law to be permanently suspended. Nor do they pray that logical laws will fail.

54. One cannot know the laws of a given world without living in it: this is why they are a posteriori. But not all knowledge of a world requires living in it—such as the general properties of laws. These are a priori. This distinction shouldn't seem surprising.

55. The "moral law"—where does this notion come from? Is it modeled on the notion of natural law, or vice versa? Both ways round would be philosophically interesting. There is, of course, nothing prescriptive about natural laws—they don't tell the world what it ought to do. A statement of natural law is not an exhortation.

56. The idea that laws exist but don't cause anything: a type of epiphenomenalism. Epiphenomenalism about natural laws: what could be more wrongheaded? The law of gravity—an idle cog! Events without laws—now they *would* be causally null. Epiphenomenalism about laws and the mind: which is more implausible? We do have a desire to cut the powerful down to size. Cartoons often make light of laws, ridiculing them almost, denying their power. Our psychological relation to laws is not irrelevant to philosophical thinking about them. (Compare our psychological relation to matter: Berkeley was surely emotionally repelled by the very idea of material entities). Perhaps we need another branch of philosophy: "affective metaphysics."

57. Could there be a nomological "zombie world"—one just like ours in the particulars it contains but lacking any laws? Could there be an "inverted world"—one that agrees with ours in its particulars but obeys different laws? (Rhetorical questions.) Then do the laws "supervene" on the particulars? Now we are headed in the wrong direction. We have become too interested in whether one thing depends on another ("co-varies"), instead of how the two are internally related. Everything supervenes on itself, after all. What if something constitutes that which it supervenes upon? At this point I am saying nothing fresh, and my feet drag. Then what about this: Could there be a world W in which the very same laws that are natural laws in our world are moral laws in W? So, for example, God has commanded everything to obey the law of gravity, with suitable rewards and punishments attached; objects, including persons, do not obey this law naturally. Oddly enough, this seems coherent. As also this: a world in which our moral laws occur as natural laws—people refrain from stealing, say, as a matter of psychological law. Would we like to live in a world in which keeping promises was part of the natural order? Try to imagine feeling guilty about not obeying the law of gravity (difficult). (Here I should insert something about the importance of *play* in philosophical thinking. Is philosophy a serious subject? Of course: but people do try to make it seem more dignified than it is, more straight-laced. Philosophical humor is not external to the subject.)

58. You can't be interested in laws without being interested in metaphysics. The name "metaphysics" is just a label for a set of problems, e.g., the problem of the nature of laws. The philosophical problem of laws is as philosophical as philosophical problems can be. It touches every philosophical nerve. It gives rise to a tingling sensation in the concept area—by turns painful and pleasant. Where do laws *fit* in the scheme of things? The characteristic philosophical question is: "What do I really think?" The coyness of our concepts, their reticence: the maddening glimpse. The marvel is that one can come to a clearer view. Surprisingly, metaphysics is possible. I often feel like I just have to take the plunge (others may feel differently). Am I being disorganized here? Not at all! Is this stream of consciousness philosophy? Well, it

does attempt to be candid, and to flow. The natural history of philosophical thought is not without interest.

59. The cause of a specific particular is actually quite general. Universals cause particulars, to put it crudely. It is not that the cause isn't specific or "local"—laws are not spread out where their instances are. A law is manifested at places *it* is not. Don't ask where the law of gravity is. The cause of a located event is not itself located. One pictures a law bearing down on its effects—or scooping them up. Particular events are the conduits of causation, rather than the mechanism of causation itself. The so-called causal power of an event is really the law it manifests. The wire is necessary for the electrical signal, but it is the signal that has the power; just so the particular is necessary for the law to operate, but the law has (*is*) the power. The law of gravity just caused the pen to drop from my hand. It would be wrong to say it was merely the opening of my fingers that did the real causing (not that events cannot be said to be causes in a derivative or secondary sense). The wide sweep of a law—yet it hooks up with the minutest particular. The cause of an event has a much broader scope than we thought. One can't help thinking of something very large causing something very small—but this is too crude. The effect of a cause is not itself a cause, in the primary sense. As it were, events float on a sea of laws. There is a fundamental asymmetry between causes and effects: laws cause events, but events don't cause laws—or indeed what laws cause. Events might be viewed as portals through which laws operate (and hence function as necessary conditions). It is not that events pay laws the courtesy of taking them along for the ride, as if *they* are the ones really in charge of the universe's journey. (Here I want to make the Wittgensteinian point that I am not denying anything the ordinary man believes.)

60. There is an analogy between laws and the mind—particularly concerning the tendency of philosophers to collapse both into their manifestations (hence "Humeanism" and behaviorism). What I would describe as independent the reductionist describes as constitutive—finding nothing more to laws and the mind than their manifestations. No doubt people see one problem through the lens of the other: behaviorism here, "Humeanism" there. No doubt I also see the problem of laws by way of the problem of mind—hence I oppose the reductive collapse. The regularity theory of laws and causation is a kind of behaviorism about laws (with or without counterfactuals). I say that events express laws as behavior expresses the mind—roughly. But we must be careful not to press the analogy too far: events cannot be *deceptively* related to laws. (The tendency to give an account of something "from a distance": we need to adopt the point of view of the thing itself.)

61. Question: If events were intrinsically imperceptible, would we still want to insist that laws reduce to their instances? Wouldn't the point of that insistence be lost? We would certainly not be making laws any less imperceptible. And what if, *per impossibile*, laws were visible and events invisible? Would we still want to reduce laws to events? Wouldn't we instead want to reduce events to laws? Would behaviorism appeal to philosophers and scientists if behavior were unobservable? Doubtful. Is

behavior *necessarily* observable? The regularity theory seems predicated on a metaphysical contingency (as does behaviorism).

62. You really feel the efficacy of a law—in your marrow, as it were—when your skis start to lose control on the downhill. Magnets provide an excellent example of hidden laws. Sometimes a law can seem to burst to the surface. The idea that the laws of nature are determined by the economy of thought is not only absurd, but also insulting (to the laws). Scoffing is characteristic of philosophers, and shows something not merely about them. Father Time, Uncle Gravity: the need to visualize. Picture the laws of nature as an august committee of ancient gentlemen. Except laws are far more elevated. We long to condense the immense into the manageable. Do we feel humiliated by laws? It feels good to cut them down to size. I am sure that some people enjoy thinking of others merely as behaving organisms too. We have no first-person perspective on laws, and so no ontological corrective. The concept of a disposition has proved a useful one to hide behind (laws as dispositions to have certain effects, mental states as dispositions to behavior). Much of philosophy is avoidance and repression.

63. There has always seemed to be something occult about causation—understandably so. Effects are given, causal powers inferred. Causation doesn't oblige the perceptual apparatus (as Hume taught us). Does regularity have anything to do with laws? Does the body have anything to do with the mind? Yes, in both cases. I would like to say that laws *animate* the world, as minds animate bodies—but that makes the relationship sound too external. The imprecision of a formulation does not always count against it. Nor is it always easy to say what counts as the literal truth. Analogies can be amazingly helpful in philosophy, as well as the source of much misunderstanding. We try to make our concepts point at each other, in the hope of mutual recognition. Concepts are reluctant to reveal their secrets in isolation (this has nothing to do with "holism"). One concept might be the key to another. I arranged a meeting between the concept of mind and the concept of law: they hit it off.

64. If we don't believe in laws, but only in particulars, doesn't causation drop out, to be replaced by mere succession? "But when we speak of causes we refer to particular events": yes, but when we speak of times we likewise refer to events—that doesn't show that times are events. After all, objects can be spoken of as causes too, but this is surely a derivative use. The question is not whether events (or facts) are truly described as causes; the question is what *account* we give of this. When does causation reach its terminus? I say laws are the last word.

65. I can easily imagine a philosopher so impressed with the reality of laws that he denies the reality of objects and events. Philosophers make a (dubious) virtue of excluding some things in favor of others. What *about* the suggestion that objects and events are merely the appearances of laws—the noumenal law and the phenomenal particular? There have been wilder ideas in philosophy. (Pan-lawism: perfectly true.) Instead of saying the *only* realities, we can say the *basic* realities. Objects and events:

"nomological danglers"? The laws are the skeleton and nervous system of the universe—with objects and events the skin.

66. There is a lot of wishful thinking in philosophy. Philosophical opinion and political opinion are close cousins. It is easy to lose one's nerve in philosophy, or to develop numbness of the nerves. I grant that my view of laws makes me nervous (and sometimes numb)—just as mathematical Platonism does. Reluctant assent is a mark of the philosophical state of mind (accompanied by a peculiar dry thrill).

67. There isn't anything *between* a particular thing and a general law, an entity of intermediate status (or "size"). Reality is either pointedly particular or majestically general (or a mixture of the two). Does anything else make sense? Nature necessarily falls into these two extremes: there are no ontological compromises—semi-universals, quasi-particulars. The reason people were driven to maintain that metaphysical questions are meaningless is that they are so obviously meaningful—and difficult to answer. I sometimes picture philosophical problems smirking silently to themselves. If philosophy just consisted of the question of laws, it would still be a big subject—because that question brings a lot with it. It sums up, encapsulates, what philosophy is. I might put the point by saying that the problem of laws is philosophy at its most stark. What *is* a law?

68. Am I trying to give a nuanced account of laws? No, I am striving for bluntness—crudity, even. Someone asks me to prove the existence of laws as I conceive them—as if I believed in ghosts or fairies. That is not the situation. Do you expect me to parade a naked law before you or take a photograph of one? Proofs of existence obviously vary with the subject matter. The obsession with whether laws are necessary or contingent. The real question is whether they are *productive*. Positivism, of course, has a lot to answer for here, in its desire to make science philosophically reputable. It may not be (by your standards). Science presupposes the efficacy of the unobservable—even when it is entirely about the observable. It is a mistake to say it is the *force* of gravity that is the true cause and not the *law* of gravity—for the force only operates as it does because of its nomological properties, i.e., the laws that govern it. If the force of gravity did not obey the inverse square law, it would not cause what it does. And note how artificial it is to try to separate the explanation of an event from its cause—as if a law could be the former but not the latter. What explains the effects of a force are the laws that govern that force. Strictly speaking, forces are no more the ultimate cause of things than events are. A cause, remember, is what makes things happen as they do—what brings about states of affairs. In a *loose* sense, even objects can be causes, as can places, times, and numbers (we use "because" very promiscuously). The question "Causes in virtue of what?" only comes to an end with the statement of law.

69. It is worth remembering that objects remain stable only because of the laws immanent in them. Stability is not the way things are *irrespective* of laws. There is never a time or place at which laws are not operating. The "Humean" tendency to think of

laws as "side-effects" of particulars—a sort of effervescence. (Compare a certain view of mind and brain.) Laws conceived as "emergent properties" of particulars. I invert this: particulars owe their very being to laws. Laws are the living tissue of nature. There is no contradiction in the notion of a law that never produces an actual regularity, since it may always be overridden by other laws. What if there was so much random wind in the universe that planets never moved in the elliptical orbits they "should" move in? The characteristic *expression* of a law is an actual regularity—but that is all.

70. Laws necessitate the *occurrence* of things: this makes their necessity differ from all other kinds, such as the mathematical kind. You cannot take a law and remove the necessity from it. The necessity is not something tacked on. We only have one good word for laws, namely "law," and it has many other meanings (legal, moral). Why this linguistic poverty? Is it because our thinking about laws is so primitive? (Compare the poverty of our modal vocabulary: we keep having to multiply the types of things we uniformly call "necessary.")

71. Physics has the best laws. Why? The natural mysticism of physicists—is it their proximity to the universe's most basic principles? How does a physicist look at the universe? As if the particular were just a symptom of a more general order. Physicists look at the world with two pairs of eyes. The particular is insignificant in physics; the laws command attention. Laws as reasons (the word is not wrongly used here): once you know the reason for things, the reason stands out from the things. Laws occupy the intellectual foreground. Maybe this causes a professional blindness in physicists, which they need to work against. (Infatuation with the universal—with the "universe.")

72. "Laws are confirmed by their instances": but after a certain point positive instances do not add further confirmation of the law. Is the law of gravity made more probable every day? No. Instances reveal laws to us; they don't aggregate on a scale of confirmation. We are inclined to picture a law as a partly concealed line the future direction of which must be predicted from its exposed segment. Then the idea of endless confirmation by instances seems appropriate. A principle can be manifest in a single instance, if the instance is good enough. A law is an "invariable sequence" (*OED*): no, it is the principle *of* such a sequence. Laws cause (bring about) invariable sequences. Do laws lurk "behind" regularities? They inform them. Laws are constant; their manifestations may not be. Whenever a cause occurs you know it will repeat itself. In *this* sense causes are uncreative.

73. We speak of laws of nature, but isn't nature defined as what is subject to laws? This gives us no independent grip on what a law is. "The concept of law is indefinable": true enough, but scarcely illuminating, since so many concepts are indefinable. Are laws "primitives"? Yes, so long as this description is not misunderstood (they are not "structureless"). Laws are not "constructions" out of anything else, least of all their manifestations. It has never really been settled what form an illuminating philosophical account should take. "To shed light on a concept": unhelpful meta-

phor. A good account of a concept should make it seem *vital*—indispensable and alive. Some concepts strike us as formidable, and rightly so.

74. Space, time, and laws: the basic ingredients of the world—any world. Without time laws would be unable to breathe. To call laws timeless is misleading: time is their natural medium. Laws and time are internally related—as objects and space are. Events occur in time, and they are the visible embodiment of laws. Since objects are constituted by laws, they are also bound up with time. It is not an accident that objects exist in time (as well as space). Laws make contact with time by means of events. An event is the occupation of time by laws. Objects, events, laws, time: a single package. The shaping of stuff by laws is essentially temporal. Objects emerge from stuff through the temporal operation of laws (this is not a statement about history). Events are therefore a pre-condition of objects: no events, no objects. Laws cannot create objects without the mediation of events (if only potentially). Neither can there be events without objects. Stuff, objects, events, and laws are coeval categories. The nexus consisting of stuff, objects, events, and laws constitutes concrete reality. Each is a facet of the others. This nexus has no natural name, but it is the crux of the world. It forms a "natural unity." One might almost say that objects, events, and laws form a "trinity." The path from laws to objects goes via events, but it is not possible to reach events without objects for laws to operate on. Objects and events are "one flesh." It is not that they are married, since that assumes a prior separation. Much of our discourse is about this nexus, but, importantly, not all (e.g., ethics, logic). There is no difficulty in *saying* things about what does not belong to the nexus. We might give a name to it, so that its unity is acknowledged: the SOLE nexus (stuff-objects-laws-events). The basis of concrete reality is SOLE. There is no deeper metaphysical category that unifies the elements of SOLE: these elements are ontologically basic. There is no further reality of which these elements are merely aspects. What is linguistically separable may not be ontologically separable. The concepts of stuff, object, law, and event don't form a "family"; they are too tightly connected for that. It is more that they constitute a single organism. Nor are they the "building blocks" of reality, since building blocks may stand alone. They are more like *perspectives* on the same reality, each containing a glimpse of the others. When you approach reality with one of these concepts you perforce bring the others along—though they are not transparently connected.

75. The connection between objects and laws is no weaker than the connection between properties and laws, since what makes an object the kind of object it is are the properties it has. The connection between properties and laws is obvious once it is pointed out. Since there are no property-less particulars, no particular can be such that it obeys no laws. The "bare particular" is not just nonsense on stilts; it is the attempted denial of the very constitutive framework of reality.

76. The *interplay* between the elements of SOLE is what is fundamental: that they are *joined* is metaphysically basic.

77. People used to talk about the "vital spirit" that inhabits all life. But isn't that idea just as appealing for objects in general? Laws are the "vital spirit" of the universe. The divide between the living and the nonliving seems less (metaphysically) pronounced once one has taken the measure of laws. Of course, the living and nonliving obey *different* laws. Equivalently: the idea of a vital spirit loses its appeal once one sees that all of nature has its own kind of "animation." Laws are intangible causes, and inexhaustible sources of power (the "entelechies" of the natural world). They have some of the properties attributed to God; yet nothing could be more "natural." It would be a complete mistake to think of them as "spiritual." Nothing is more mindless. A law has no awareness of itself.

78. The relation between a law and its manifestations is like nothing else. It is hard to find the right words for this relation. The law does not consist in its manifestations, and it does not "underlie" them. Trying to understand this relation is like trying to squeeze between two concepts that are stuck together.

79. The characteristic condition of the philosopher is embarrassment.

80. We must try to speak of what we want badly to pass over in silence.

Index

A
activity, 203–4
affective metaphysics, 224
agency, 203
　human, 215
　of natural laws, 215
aging, 120–21
air, 111n2
analogies, 226
The Analysis of Matter (Russell), 24n13, 32n22, 144
The Analysis of Sensations (Mach), 9n9
angels, 30
animals, 11n14
　perception of, 148n8
　purpose in, 152
antimechanism, 205
antirealism, 156, 156n22
aphorisms, 219, 222
Aristotle, 111, 118, 119, 120, 182, 184
atheism, 223
atomism, 50, 194
　Displacement Theory and, 107
　in geometry, 197
　impenetrability in, 193n1
　mathematics and, 196n9
　meaning in, 196, 198, 199, 200, 204
　mind and, 208
　physics from, 192
　size in, 53
　unification in, 52
atoms
　divisibility of, 55, 56
　size of, 56, 100n4
Austin, J. L., 202

B
beauty, 214
behavior, as motion, 206
behaviorism, 95, 205–6, 206n22, 225
Berkeley, George, 32n23, 224
big bang, 34, 39, 51, 181, 181n6
biology
　constancy of, 120–21
　definition of, 142, 143
　epistemology of, 152–54
　function in, 152–53, 153n17
　molecular, 153–54
　motion in, 98–99, 99n3, 107
　physics and, 151–52, 153–54
　psychology and, 151–52, 153
black holes, 48, 71, 125
black holes, super-, 39, 39n29
body
　aging of, 120–21
　mind and, 31, 61, 86n14, 103
　as mystery, 154, 157n24
Boscovich, Roger Joseph, 193, 193n1
brain
　consciousness and, 103n7
　motion in, 102–3
　particles of, 99n3, 102–3
　physical, 154
breakability, 55
Bridgman, Percy Williams, 205

C
Carroll, Lewis, 143
Cartesian materialism, 175, 177, 177n1, 179, 189
Cartesian mechanism, 9, 193
causal complexity, 66

causation
    definition of, 213
    through events, 225
    Hume on, 128
    as illusion, 170n6
    as infinite, 215
    from natural laws, 215, 218–19, 226, 228
    through objects, 227
    as occult, 226
    in psychology, 220
    remote, 10
    repetition from, 218
change
    color, 99, 101, 105n10
    in consciousness, 99n3, 121n8
    conservativeness of, 105, 106
    definition of, 98n2, 109n14, 109n15
    of direction, 110, 119
    Displacement Theory of, 98, 104, 105, 107, 109
    endogenous force for, 114, 115, 115n5
    intrinsic, 104, 117
    in light, 106
    macroscopic, 99–100, 101n4
    of matter, 115n6
    in meaning, 205
    microscopic, 100, 100n4
    of mind, 101n5, 102, 103, 103n8
    from motion, 98, 98n2, 99–101, 104–5, 105n10, 106, 107
    in motion, 108–9, 109n14, 109n15, 111, 112–14, 114n4, 115, 115n5, 116–17, 119–20
    of place, 104, 105n10, 116–17
    relational, 105, 117
    in semantics, 204
    of shape, 104, 105n10, 107
    of size, 105n10
    space and, 101–2, 102n6
chaos, 221
charge. See electricity
chemistry, 198n8
Chomsky, Noam, 7n6, 8n9
collision, 33
color, 99, 101, 105n10
concepts, 163, 163n30, 164, 206–7, 226
conceptual holism, 207n24
consciousness, 6–8, 8n8. See also psychology
    brain and, 103n7
    change in, 99n3, 121n8
    constancy of, 121
    definition of, 64, 187–88
    emergence of, 184, 185, 189
    end of, 180n5
    as energy, 174n10, 178, 180, 180n5
    as force field, 31n20
    matter as, 61–62, 62n8, 67–68, 72, 93, 161–62, 162n28, 178, 179, 180, 180n4, 181–82, 182n7, 183–85, 185n12, 186, 186n13, 187, 188, 188n16, 189, 190, 191
    matter's effect on, 146n7
    mind and, 70n15
    as mystery, 157, 157n23, 158, 189
    origins of, 181, 189
    in physics, 188n15
    size of, 72
    subjectivity of, 150n12
    types of, 52
    unconsciousness, 64
    volition and, 63
conservation
    of energy, 166, 166n1, 167–68, 171, 172–73, 180, 204n20
    of meaning, 204n20
    with semantics, 204n20
    of solidity, 35, 35n25
    of space, 34, 35
constancy
    of biology, 120–21
    of consciousness, 121
    of mass, 113, 115
    of matter, 35, 115n6, 204
    of motion, 111, 112–15, 115n5, 116–17, 118, 171
    of natural laws, 215, 219, 220, 221–22, 227–28
    as norm, 115–16, 120–22
    of rest, 118
    of semantics, 204n20
    of shape, 113–14
conventionalism, 5
cutting, 55, 55n2

D
dark matter, 177
death, 120–21
deletion, 18

Deletion Theory, 42
 gravity in, 39–40
 impenetrability in, 29n18
 matter in, 46, 49
 motion in, 34–35
 particle size in, 54
 space in, 19, 19n8, 20, 21–24, 24n12, 25–29, 30, 31, 33, 34–35
density, 48
Descartes, René, 8, 15–16
 Discourse on Method, 12
 dualism of, 49n2, 178n2, 179, 179n3, 180, 182, 184, 184n10
 on extension, 12, 13, 14, 192–93
 materialism of, 175, 177, 177n1, 179, 189
 on matter, 17, 192–93, 199
 mechanism of, 193
 Meditations and Other Metaphysical Writings, 12–13
 on mind, 29, 30, 101n5, 103, 103n8
 The Principles of Philosophy, 12
Dilution Theory, 46
Dirac, Paul, 24n12
direction, 77, 110, 119
Discourse on Method (Descartes), 12
displacement, 139, 140
Displacement Theory, of change, 98, 104, 105, 107, 109
distance, 70–71
divisibility
 of atoms, 55, 56
 cutting, 55, 55n2
 functional, 55–56
 geometric, 54–55
 indivisibility, 53, 54–55
 of matter, 56–57
 of particles, 53, 54–56
drive, 205–6
dualism, 49n2, 178n2, 179, 179n3, 180, 182, 184, 184n10
Dummett, Michael, 3–4
duration, 84
dynamic universe, 22
dynamism, 193–94, 203–4

E
earth
 magnetic field of, 16
 as stationary, 82
ectoplasm, 60n2

Eddington, Arthur, 5, 61, 62n8, 143, 143n2, 144, 146, 150
Einstein, Albert, 78n6, 89, 90, 165, 194
electricity, 10. *See also* electromagnetism
 behavior of, 155
 cause of, 134, 135
 charge, 131–33, 133n4, 134, 135, 136, 137, 137n9, 139–40, 141n12, 153, 155, 166, 168n5, 193
 cohesion from, 198n8
 definition of, 94, 128–30, 136, 136n8, 137
 epistemology of, 153
 etymology of, 130–31, 131n2
 as fluid, 136n8
 as force field, 137, 138
 impenetrability of, 129n1
 motion of, 132n3
 as mystery, 135n7, 137, 140–41, 141n12, 154n19
 neutral, 134, 140
 ontology of, 131
 repulsion of, 43
 resinous, 134n5
 sensation of, 87n15, 139
 vitreous, 134n5
electrodynamics, 139
electromagnetism
 gravity and, 134, 135, 136
 motion and, 92, 97
 theory of, 90
electrons
 effects of, 202n17
 mass of, 49
 motion of, 104n9
 shape of, 93
electrostatics, 139
elements, 50–51
Elements (Euclid), 197
eliminativists, 189
$e=mc^2$, 165
emergence, 36–37, 184, 185, 189
empirical equivalence, 29n18
empiricism, 62, 62n8, 207n24, 214
Encyclopedia of Philosophy (Jammer), 96
energy
 as autonomous, 194
 boundaries of, 201
 consciousness as, 174n10, 178, 180, 180n5

energy (*continued*)
  conservation of, 166, 166n1, 167–68, 171, 172–73, 180, 204n20
  definition of, 64, 64n9, 165, 166–67, 169–70
  intrinsic, 170, 171
  matter as, 63–64, 194, 194n3, 200, 203
  measurement of, 166, 172
  of mind, 174n10, 180
  as mystery, 174
  particles as, 194
  in quantum theory, 194n2, 194n3
  radiant, 60n5
  as substance, 194n2, 194n3
  in systems, 173n9
  transference of, 64n9, 168, 168n4, 171–73
  transmutability of, 10
  types of, 169, 172n8
  uniformity of, 173–74
entities, 214
epiphenomenalism, 224
Epstein, Lewis Carroll, 145n6
An Essay Concerning Human Understanding (Locke), 14, 68, 68n12
ether, 85, 160, 175, 177
ethics, 213–14
Euclid, 27n15, 197
events, 204, 204n19
  causation through, 225
  explanation of, 227
  natural laws of, 219, 222
  objects in unity with, 229
  perception of, 225
  in time, 226, 229
evil, 214
evolution, 104
exclusion, 16, 16n6
  by matter, 17–18, 19, 23–24
  of mind, 30–31
  in quantum theory, 36
  by space, 20–21
experience, 106n11
extension
  belonging of, 20n9
  Descartes on, 12, 13, 14, 192–93
  impenetrability and, 24–25
  macro-, 37n27
  mass and, 50
  matter as, 23, 58–59, 199

  of objects, 24, 40, 41
  proper, 40, 41
  in space, 15–16
  of space, 22–23

F
fields, force, 175
  boundaries of, 202n16
  center of, 45–46
  consciousness as, 31n20
  electricity as, 137, 138
  gravity, 16, 39–40
  impenetrability from, 43
  as internal, 23
  magnetic, 16, 17, 21
  matter as, 23–24, 36, 176
  mind and, 31, 207
  overlap of, 43, 46
  particles as, 44–45, 193–94
  in realism, 138
  resistance in, 45–46
  space and, 102
force
  for change, 114, 115, 115n5
  as essential, 194
  meaning as, 199–200, 203
  without motion, 91
  relative, 80, 80n8
  as secondary, 193
  sense as, 199–200
Forms, of Plato, 215, 222
Franklin, Benjamin, 131
Freedman, Roger A., 129, 129n1, 130–31, 137
Frege, Gottlob, 197–98, 198n8, 201
friction, 111n2, 112, 118
functionalism, 94, 149n10, 206

G
Galileo Galilei, 175
gases, 48
Geach, Peter, 79n7
General Chemistry (Pauling), 60n5
General Relativity Theory, 24n12, 40, 72, 102n6
genetics, 153–54
geometry, 196, 197n7
  atomism in, 197
  boundaries in, 202n16
  divisibility in, 54–55

God, 5n4, 26, 27, 27n15
　definition of, 64–65, 65n11
　matter as, 64–65, 65n11
　motion of, 75, 75n2
　as mystery, 190n17
　natural laws and, 214, 215, 217, 219, 220, 221
　sensation of, 219
　substance of, 214
Goethe, Johann Wolfgang von, 203, 205
gravity, 5n4, 10
　curvature of, 24n12
　definition of, 40–41, 70–71
　in Deletion Theory, 39–40
　distance and, 70–71
　electromagnetism and, 134, 135, 136
　force field of, 16, 39–40
　ignorance of, 72, 72n16
　impenetrability as limit of, 38–39
　inertia and, 124–25, 127
　law of, 125–27
　mass and, 39, 41, 70–72, 123–24, 126, 136
　matter and, 39–40, 72–73
　motion and, 92, 126, 127
　as mystery, 190
　Newton on, 72n16, 124n1
　as occult, 175
　place of, 225
　source of, 39–40, 124n1
　space and, 40
Greene, Brian R., 76–77
Grice, Paul, 202
growth, 98, 101, 103, 107, 120

H
heat, 97, 100, 106, 106n11
heaven, 216
Heaviside, Oliver, 167
Heisenberg, Werner, 49, 180n4, 194n2, 194n3, 196n10
"Hempel's Dilemma," 8n10
Heraclitus, 120, 194n2, 203
Hertz, Heinrich, 5, 5n2, 9
History of England (Hume), 7n6
holism, 206–7, 207n24
How to Do Things With Words (Austin), 202
humans, 11n14
Hume, David, 7n6, 128, 223

I
idealism, 32, 32n23, 62, 159n26
ideas
　impenetrability of, 32n23
　laws and, 214
　in mind, 177n1
identity, 44
　of mind, 61
　of place, 79, 79n7, 80
　relative, 79n7
identity theory, 183
impenetrability
　in atomism, 193n1
　as constant, 35
　in Deletion Theory, 29n18
　of electricity, 129n1
　extension and, 24–25
　from force fields, 43
　as gravity limit, 38–39
　of ideas, 32n23
　of matter, 13, 14, 15, 16, 16n6, 17, 18, 19, 24, 37, 68n12
　of mind, 30, 31
　motion and, 33, 34
　of objects, 13, 24
　Overlap Theory and, 29n18
　in quantum theory, 37n27
　of space, 18–19, 26, 35n26, 46–47
　strong, 43–44, 44n1, 45
　weak, 43–44
indivisibility, 53, 54–55
inertia, 38, 63, 71
　gravity and, 124–25, 127
　law of, 106n11, 111, 112, 115, 115n6, 116–21, 121n8, 122, 122n9
　mass and, 123, 125, 126
　of particles, 49
　of planets, 125
instantiation, 121
instrumentalism, 5, 159n26
intelligence, 11n14
intentionality, 31
intermittent entity, 22, 22n10
intimate knowledge, 143
introspection, 148n9, 206n23
isolation, 121–22
italics, 217

J

"Jabberwocky" (Carroll), 143
Jammer, Max, 96
Journal of Philosophy, 7n6

K

Kant, Immanuel, 69n13, 106n11
knowledge
  by inference, 147
  intimate, 143
  from introspection, 148n9, 206n23
  limits of, 191n18
  of natural laws, 218, 219, 222, 223
  remote, 143, 146
  scientific, 11n14
  symbolic, 146

L

Lange, Marc, 138n10, 167
language, 150, 150n13, 151, 151n14
  as dynamic, 199
  as natural laws, 220
  sentence, 195, 196n6
law, moral, 224
laws, legal, 222
laws, natural, 213
  abrogation of, 217
  acceptance of, 221
  agency of, 215
  awe of, 219–20
  causation from, 215, 218–19, 226, 228
  conditions for, 216
  constancy of, 215, 219, 220, 221–22, 227–28
  counterfactuals and, 223
  definition of, 228–29
  discovery of, 218, 219–20, 221
  epiphenomenalism about, 224
  of events, 219, 222
  God and, 214, 215, 217, 219, 220, 221
  human relationship with, 222, 226
  ineffability of, 223
  of inertia, 106n11, 111, 112, 115, 115n6, 116–21, 121n8, 122, 122n9
  as infinite, 215, 218, 219, 222
  knowledge of, 218, 219, 222, 223
  as language, 220
  mind and, 221, 225
  of motion, 71, 110, 111, 112, 112n3, 115, 117, 119– 120, 193, 216, 220
  objects and, 213n2, 214, 217, 227, 229
  as occult, 218
  perception of, 222, 225–26
  in philosophy, 224–25, 227
  place of, 225
  power of, 220, 221, 225, 230
  proof of, 227
  of psychology, 216, 220
  regularities of, 218, 226, 227, 228
  sensation of, 216, 221
  as side effect, 227–28
  simplicity of, 222
  size of, 220
  in time, 215, 229
  as transcendent, 215
  as uniform, 223
  world without, 224
Leibniz law argument, 23n11, 79, 79n7
length, 138n11
light
  change in, 106
  matter and, 177
  shape of, 60n2
  speed of, 10, 60n5, 89, 89n18, 90–91, 101, 177
Locke, John, 15, 18, 19, 30
  An Essay Concerning Human Understanding, 14, 68, 68n12
  "Of Solidity," 14
Lockwood, Michael, 145n6
logic, 197n7, 214, 215
logical space, 196–97, 197n7

M

Mach, Ernst, 9n9, 62n8
magnetic fields, 16, 17, 21
magnets, 226
Malcolm, Norman, 195
mass
  constancy of, 113, 115
  definition of, 37–38, 48, 63, 71, 126, 205, 206
  of electrons, 49
  extension and, 50
  gravity and, 39, 41, 71–72, 123–24, 126, 136
  inertia and, 123, 125, 126
  matter as, 63
  as measure, 205
  motion and, 93, 111, 126–27
  of neutrons, 49

of objects, 48, 49–50, 71
of particles, 49, 50, 52, 53
of planets, 82n10
of protons, 49
relative, 80
size and, 48–49
volume and, 49–50
zero-, 63
material, 185n11, 186n13
materialism, 8, 160, 182n8, 185–86. *See also Real Materialism and Other Essays*
 antimaterialist, 187
 Cartesian, 175, 177, 177n1, 179, 189
 pan-, 182
mathematics, 42n31, 89
 atomism and, 196n9
 from logic, 215
 motion in, 88, 88n17
 physics and, 10–11, 11n14, 88n17, 144n3, 153, 153n18
matter
 as agent of motion, 104
 as annihilating space, 18–19, 20, 21–22, 25, 102n6
 from big bang, 39, 51, 181, 181n6
 change of, 115n6
 as consciousness, 61–62, 62n8, 67–68, 72, 93, 161–62, 162n28, 178, 179, 180, 180n4, 181–82, 182n7, 183–85, 185n12, 186, 186n13, 187, 188, 188n16, 189, 190, 191
 constancy of, 35, 115n6, 204
 continuity of, 26, 50
 creation of, 27
 dark, 177
 definition of, 12–15, 16, 16n6, 17, 38, 38n28, 48, 58–59, 60, 60n5, 61–63, 66–67, 68–69, 93, 94–95, 177, 179n3, 186, 188n15
 in Deletion Theory, 46, 49
 density of, 48
 as discrete, 107
 disunity of, 50–52
 divisibility of, 56–57
 effect of, on consciousness, 146n7
 as energy, 63–64, 194, 194n3, 200, 203
 exclusion by, 17–18, 19, 23–24
 as extension, 29, 58–59, 199
 as force field, 23–24, 36, 176
 as God, 64–65, 65n11
 gravity and, 39–40, 72–73
 ignorance of, 67–68, 68n12, 69, 69n13, 70, 70n15
 impenetrability of, 14, 15, 16, 16n6, 17, 18, 19, 24, 37, 68n12
 as inert, 203
 light and, 177
 as mass, 63
 mind as, 30–31, 32, 157n24, 179n3, 213
 as mystery, 7, 8n8, 157, 158, 189–90
 negative, 24n13
 Newton on, 69n13
 origins of, 39
 perception of, 150n12
 as phenomenal, 63
 positive, 24n13
 psychology of, 162n28
 in quantum theory, 36, 37n27
 resistance of, 17–18, 68n12
 shape of, 59–60
 solidity of, 21, 36–37, 59
 space and, 23n11, 95n24, 214
 structure of, 192, 193
 types of, 50–52, 178, 178n2, 179, 184, 188n15
 as unconsciousness, 64
 unity of, 49–51, 52, 180n4
 as volition, 63
 zero-mass, 63
Maxwell, Clerk, James 9, 64n9, 76n5
meaning, 150, 151
 in atomism, 196, 198, 199, 200, 204
 boundaries with, 201
 change in, 205
 conservation of, 204n20
 as force, 199–200, 203
 as homogeneous, 196
 overlap in, 204n19
 as passive, 203, 204
 as potential, 201n14, 204–5
 semantics and, 201
 of sentence, 199
 sentence radical, 200, 200n13
 structure of, 196, 197, 199n11, 201, 202
 as substance, 195
 through will, 205
measurement, 138n11, 146n7
mechanism, 8, 175, 183n9
 anti-, 205
 Cartesian, 9, 193
 in semantics, 205

Meditations and Other Metaphysical Writings (Descartes), 12–13
mentalism, 32
metaphors, 207, 207n25, 208, 219
metaphysics, 4, 223, 224
microphysics, 194
mind. *See also* brain; consciousness; meaning; will
  atomism and, 208
  body and, 31, 61, 86n14, 103
  change of, 101n5, 102, 103, 103n8
  consciousness and, 70n15
  definition of, 29–30
  energy of, 174n10, 180
  exclusion by, 30–31
  force field and, 31, 207
  ideas in, 177n1
  identity of, 61
  impenetrability of, 30, 31
  as matter, 30–31, 32, 157n24, 179n3, 213
  mental substance of, 49n2
  mystery of, 157n24
  natural laws and, 221, 225
  overlap of, 44n1
  philosophy of, 6
  self-awareness with, 158
miracles, 217
molecular biology, 153–54
moral law, 224
More, Henry, 13, 30
motion, 6
  absence of, 92
  absolute, 74, 75, 77–78, 80, 81, 82, 82n9, 83, 83n11, 84–85, 85n13, 86–87, 87n15, 89n18, 90, 91, 109n15, 118, 140
  annihilation as, 107–8, 108n12
  apparent, 82n10
  Aristotle on, 111
  behavior as, 206
  in biology, 98–99, 99n3, 107
  in brain, 102–3
  change from, 98, 98n2, 99–101, 104–5, 105n10, 106, 107
  change in, 108–9, 109n14, 109n15, 111, 112–14, 114n4, 115, 115n5, 116–17, 119–20
  constancy of, 111, 112–15, 115n5, 116–17, 118, 171
  definition of, 15, 33, 74, 78, 87, 94n23, 108–9
  in Deletion Theory, 34–35
  electrical, 132n3
  electromagnetism and, 92, 97
  of electrons, 104n9
  force without, 91
  of God, 75, 75n2
  gravity and, 92, 126, 127
  heat as, 97, 100, 106, 106n11
  illusion of, 78
  impenetrability and, 33, 34
  mass and, 93, 111, 126–27
  in mathematics, 88, 88n17
  matter as agent of, 104
  mobility, 127
  motive power, 127
  as mystery, 108–9
  natural laws of, 71, 110, 111, 112, 112n3, 115, 117, 119–120, 193, 216, 220
  Newton's first law of, 110, 111, 114, 117, 119–20
  objects in, 27
  over time, 101, 126
  perception of, 75–77, 78, 79, 81, 82, 82n9, 82n10, 87, 87n15, 88, 94n23, 109n14, 111
  in physics, 91, 92, 92n20, 93–94, 95, 96–97, 104
  of planets, 80, 118–19, 127
  real, 81, 91, 122
  relative, 75–79, 80, 82, 83, 83n11, 85n13, 86–87, 89n18, 91, 95, 118
  shape and, 98
  shape of, 118–19
  size and, 98, 126n1
  speed of, 101
  thoughts as, 206n23
  uniform, 122
motion functionalism, 94
motivation, 205
movement, 27
"Mysteries of Nature: How Deeply Hidden?" (Chomsky), 7n6

N
natural laws. *See* laws, natural
natural metaphysics, 4
natural philosophy, 3

nature, 228
The Nature and Future of Philosophy
  (Dummett), 3–4
The Nature of the Physical World
  (Eddington), 143, 143n2, 144
neurons, 154
neuroscience, 154
neutrons, 49, 140, 196n10
Newton, Isaac, 7n6, 71
  first law of motion, 110, 111, 114, 117, 119–20
  on gravity, 72n16, 124n1
  influence of, 198
  laws of motion, 110, 111, 112, 114, 115, 117, 119–20, 193, 220
  on matter, 69n13
  Principia, 5n4, 165, 197
  on space, 74–75, 77
Newtonian mechanics, 6–7, 8, 9
nomologicism, 215
nothingness, 28–29, 65–66
numbers, 101

O
objectivity, 62, 155n20
objects
  causation through, 227
  chaos of, 221
  counting, 43–44
  events in unity with, 229
  extension of, 24, 40, 41
  impenetrability of, 13, 24
  individuation of, 43–44, 44n1, 45
  indivisibility of, 54–55
  mass of, 48, 49–50, 71
  in motion, 27
  natural laws and, 213n2, 214, 217, 227, 229
  physicality of, 59, 59n1
  primary qualities of, 58–59, 61
  reality of, 215
  at rest, 110
  secondary qualities of, 58, 61
  shape of, 59–60
  size of, 49, 53–54
  space as, 21–22
  substance of, 58, 213
  supernatural, 214
  in time, 215
  weight of, 111n2

observation. *See* perception
"Of Solidity" (Locke), 14
overlap
  of force fields, 43, 46
  in meaning, 204n19
  of particles, 44n1
  of space, 44n1
Overlap Theory, 20, 20n9, 22–23, 23n11, 27
  impenetrability and, 29n18
  space in, 21, 28, 34–35

P
pain, 69, 160, 183n9
panmaterialism, 182
panpsychism, 52, 52n6, 61–62, 62n8, 63, 67–68, 72, 157n24, 162n28, 182, 182n8
pantheism, 64–65
particles, 26, 38, 38n27
  boundaries of, 201
  of brain, 99n3, 102–3
  composition of, 49, 180n4
  cutting, 55, 55n2
  definition of, 196n10
  divisibility of, 53, 54–56
  as energy, 194
  as force field, 44–45, 193–94
  indivisible, 53
  inertia of, 49
  large, 53, 53n1, 57, 57n3
  mass of, 49, 50, 52, 53
  overlap of, 44n1
  point, 54, 55, 56
  in quantum theory, 44–45
  radius of, 50
  shape of, 93
  size of, 53, 53n1, 54, 56, 57, 57n3
  substance of, 60
passivity, 203–4
Pauling, Linus, 60n5
perceptibility, 17
perception
  animal, 148n8
  of events, 225
  of matter, 150n12
  of motion, 75–77, 78, 79, 81, 82, 82n9, 82n10, 87, 87n15, 88, 94n23, 109n14, 111
  of natural laws, 222, 225–26
  physics and, 147–48, 150
  in psychology, 148–49, 149n10, 150, 155

percepts, 160n27
perseverance, law of, 115–16
persistence, 205
Philosophical Investigations (Wittgenstein), 198–99, 200, 201
philosophy
  concepts in, 163–64
  definition of, 162, 163, 164
  emphasis in, 217
  importance of, 164n31
  insight in, 223
  of mind, 6
  natural, 3
  natural laws in, 224–25, 227
photons, 97, 108, 108n12
physical, definition of, 8n10, 9, 31–32, 184n10
physicalism, 8, 9–10, 32n21, 182n8, 222
physics
  abstractness of, 10–11, 147–48, 160n27
  antirealist interpretation of, 156
  from atomism, 192
  biology and, 151–52, 153–54
  consciousness in, 188n15
  definition of, 142–43
  disunity of, 8
  Eddington on, 143–44, 146
  envy of, 11
  epistemology of, 64n9
  as functionalism, 206
  influence of, 81, 192, 193, 195–97, 198n8, 198n9, 200, 201–2, 202n16, 202n18, 203–4, 204n19, 205–6, 207, 208
  mathematics and, 10–11, 11n14, 88n17, 144n3, 153, 153n18
  mechanics of, 6, 146–47
  meta-, 42
  metaphysics, 4, 223, 224
  motion in, 91, 92, 92n20, 93–94, 95, 96–97, 104
  as operationalist, 41, 70, 81, 94, 95, 103–4, 138n11, 143–44, 153, 168, 168n5, 202n18, 205
  paradox of, 155
  perception and, 147–48, 150
  realism in, 156, 156n22, 161
  reductionism to, 159, 177n1
  selectivity of, 7n7
  shape in, 93
place
  change of, 104, 105n10, 116–17
  of gravity, 225
  identity of, 79, 79n7, 80
  of natural laws, 225
  observation of, 75–77
  relative, 109n14
  in relativism, 78–79
planets, 16, 71, 72, 82
  inertia of, 125
  mass of, 82n10
  motion of, 80, 118–19, 127
Plato's Forms, 215
plenum, 13, 14
Poincare, Henri, 5
  on laws of motion, 112, 112n3
  Science and Hypothesis, 144, 144n4, 145
politics, 214
position. *See* place
positivism, 5–6, 9, 78, 78n6, 90, 159n26, 227
positrons, 24n12
potentia, 194n3, 201n14
potential, 201n14
Principia (Newton), 5n4, 165, 197
Principia Mathematica (Russell and Whitehead), 197
"Principle of Relativity," 76
The Principles of Mechanics (Hertz), 5n2
The Principles of Philosophy (Descartes), 12
probability wave, 194n3
The Problems of Philosophy (Russell), 92n20
propositions, 197, 204n19
protons, 49, 202n17
psychology
  behaviorism, 95, 205–6, 206n22, 225
  biology and, 151–52, 153
  causation in, 220
  definition of, 142, 143
  interpretation in, 155–56
  of matter, 162n28
  measurement in, 206n22
  methodological restrictions of, 95n25
  natural laws of, 216, 220
  objectivity in, 155n20
  perception in, 148–49, 149n10, 150, 155
  realism in, 156
purpose, 152

Q
quantum indeterminacy, 202
quantum theory, 4, 155
  duality in, 184
  energy in, 194n2, 194n3
  exclusion in, 36
  impenetrability in, 37n27
  interpretation of, 156, 156n21
  matter in, 36, 37n27
  as mystery, 190
  particles in, 44–45
  solidity in, 37n27
quarks, 53–54
Quine, Willard Van Orman, 202, 202n18

R
radiant energy, 60n5
radius, 50
Ramsey, Frank, P., 145n5
realism, 138, 145n6, 156, 156n22, 161
reality
  non-spatial, 28
  as temporal, 215
Real Materialism and Other Essays
  (Strawson), 31n20, 145n6
reason, 106n11
reductionism, 98, 159, 225
regularities, 218
relationalism, 145
relative identity, 79n7
relativism, 76, 76n5, 79n7
Relativity Visualized (Epstein), 145n6
remote knowledge, 143, 146
repletion, 15
resistance, 15, 16n6
  in force fields, 45–46
  of matter, 17–18, 68n12
rest, constancy of, 118
Russell, Bertrand, 5, 32, 61, 160n27
  The Analysis of Matter, 24n13,
    32n22, 144
  Principia Mathematica, 197
  The Problems of Philosophy, 92n20
  on substance, 204n19
  theory of descriptions, 196

S
Schopenhauer, Arthur, 62–63
Science and Hypothesis (Poincare), 144,
  144n4, 145

science fiction, 222
scientific knowledge, 11n14
semantic indeterminacy theory, 202
semantics, 150, 196
  change in, 204
  conservation with, 204n20
  constancy of, 204n20
  holism in, 206–7, 207n24
  meaning and, 201
  mechanism in, 205
  semantic indeterminacy theory, 202
  vagueness in, 201–2
sensation, 64n9
  of electricity, 87n15, 139
  of God, 219
  of natural laws, 216, 221
sense, theory of, 197–98, 198n8,
  199–200
sentence, 195, 196n6
  as dynamic, 199
  meaning of, 199
sentence radical, 200, 200n13
shape
  change of, 104, 105n10, 107
  constancy of, 113–14
  of electrons, 93
  of light, 60n2
  of matter, 59–60
  of motion, 118–19
  motion and, 98
  of objects, 59–60
  of particles, 93
  in physics, 93
  of space, 59, 60n2
simultaneity, 78n6
size
  in atomism, 53
  of atoms, 56, 100n4
  change of, 105n10
  of consciousness, 72
  as contingent, 57n3
  mass and, 48–49
  motion and, 98, 126n1
  of natural laws, 220
  of objects, 49, 53–54
  of particles, 53, 53n1, 54, 56,
    57, 57n3
  as relative, 54
sleep, 22
SOLE. *See* stuff-objects-laws-events

solidity, 14, 15, 19
   conservation of, 35, 35n25
   mass and, 38
   of matter, 21, 36–37, 59
   in quantum theory, 37n27
   as respectability, 221
space
   as absence, 25–26, 27, 27n15, 28, 40
   absolute, 65n5, 75, 82n9, 86n14
   as allowing movement, 27
   boundary of, 35, 35n26, 36, 83–84, 86
   change and, 101–2, 102n6
   conservation of, 34, 35
   as constant, 35
   continuity of, 26
   coordinates in, 85–86, 102n6
   curvature of, 24n12
   definition of, 13, 14–15, 69n13
   in Deletion Theory, 19, 19n8, 20, 21–24, 24n12, 25–29, 30, 31, 33, 34–35
   in Dilution Theory, 46
   as Euclidian, 27n15
   extension in, 15–16
   extension of, 22–23
   extra dimension in, 41
   flexibility of, 34
   force fields and, 102
   full, 20, 20n9, 26–27
   gravity and, 40
   holes in, 24n12, 25, 26–27, 39, 40, 40n30, 41
   as immovable, 33
   impenetrability of, 18–19, 26, 35n26, 46–47
   as infinite, 83
   internal, 19, 19n8, 20–21, 22, 23, 28, 41–42
   mathematization of, 42n31, 89
   matter and, 23n11, 95n24, 214
   matter as annihilating, 18–19, 20, 21–22, 25, 102n6
   measurement of, 77
   motion of, 75, 75n2
   Newtonian, 74–75
   as nothingness, 28–29, 65–66
   as object, 21–22
   occupancy in, 15, 16–17, 18, 19, 21, 66, 83
   overlap of, 44n1
   receptivity of, 20, 20n9, 21, 33
   shape of, 59, 60n2
   storage in, 34
   structure of, 24, 79, 83
   uniformity of, 81
   as vacuum, 14, 15
spatio-temporal location, 44, 44n1, 45
Special Relativity, 89
static universe, 22
storage, 34
strangeness, 19–20
Strawson, Galen, 31n20, 145n6, 148n9, 182n8
strings, 56
String Theory, 41, 55, 93–94
structuralism, 4–5, 5n2, 6, 10, 87, 87n15, 122, 145–46, 148, 157
structural realism, 145n6, 156
stuff-objects-laws-events (SOLE), 229
subjectivism, 62n8
subjectivity, 62, 62n8
substance, 60
substantialism, 20n9
substratum, 213
super-black holes, 39, 39n29
supernatural, 120, 214, 217
supervenience, 99n3, 102, 173n9, 187, 224
survival, 153
symbolic knowledge, 146
syntax, 150

T
temperature, 97, 100, 106, 106n11
theism, 223
thoughts, 177n1, 182, 206n23
tides, 97
time
   absolute, 84
   events in, 226, 229
   mathematization of, 88n17
   motion over, 101, 126
   natural laws in, 215, 229
   objects in, 215
Tractatus (Wittgenstein), 195, 196–97, 199

U
undulations, 207
unity, of matter, 49–51, 52, 180n4
University Physics (Young and Freedman), 129, 129n1, 130–31, 137

V

vagueness, 201–2
velocity, 110
verificationism, 9–10, 78n6, 83, 86n14, 207n24
vision, 109n14
volition, 63
volume, 48, 49–50

W

water, 56
wave, 193, 207
weight, 111n2

Whitehead, Alfred North, 197
Wiggins, David, 79n7
Wilde, Oscar, 222
will, 62–63, 174, 204, 205
Wittgenstein, Ludwig, 195, 195n5, 196–97, 198–99, 200, 201, 204n19
Worrall, John, 145n6

Y

Young, Hugh D., 129, 129n1, 130–31, 137

Z

zero-mass, 63

www.ingramcontent.com/pod-product-compliance
Ingram Content Group UK Ltd.
Pitfield, Milton Keynes, MK11 3LW, UK
UKHW022153230426
12049UKWH00003BA/73